Robin Pfefferkorn

EAS Elements for Solid Mechanics

Mesh Distortion Insensitive and Hourglassing-Free Formulations with Increased Robustness

Schriftenreihe des Instituts für Mechanik
Karlsruher Institut für Technologie (KIT)

Band 10

Herausgeber:
Prof. Dr.-Ing. habil. Peter Betsch
Prof. Dr.-Ing. habil. Thomas Seelig

Eine Übersicht aller bisher in dieser Schriftenreihe erschienenen Bände finden Sie am Ende des Buches.

EAS Elements for Solid Mechanics

Mesh Distortion Insensitive and Hourglassing-Free Formulations with Increased Robustness

by
Robin Pfefferkorn

KIT Scientific Publishing

Karlsruher Institut für Technologie
Institut für Mechanik

EAS Elements for Solid Mechanics – Mesh Distortion Insensitive and Hourglassing-Free Formulations with Increased Robustness

Zur Erlangung des akademischen Grades eines Doktor-Ingenieurs von der KIT-Fakultät für Bauingenieur-, Geo- und Umweltwissenschaften des Karlsruher Instituts für Technologie (KIT) genehmigte Dissertation

von Robin Pfefferkorn, M.Sc.

Tag der mündlichen Prüfung: 28. Oktober 2022
Hauptreferent: Prof. Dr.-Ing. habil. Peter Betsch
Korreferent: Prof. Dr.-Ing. habil. Sven Klinkel

Impressum

KIT Scientific Publishing

Karlsruher Institut für Technologie (KIT)
KIT Scientific Publishing
Straße am Forum 2
D-76131 Karlsruhe

KIT Scientific Publishing is a registered trademark of Karlsruhe Institute of Technology.
Reprint using the book cover is not allowed.

www.ksp.kit.edu

Print on Demand 2023 – Gedruckt auf FSC-zertifiziertem Papier

ISSN 2363-4936
ISBN 978-3-7315-1271-4
DOI 10.5445/KSP/1000154549

Abstract

Proposed in the early 1990s, the *enhanced assumed strain* (EAS) method is one of the probably most successful mixed finite element methods for solid mechanics. The present cumulative dissertation gives a comprehensive overview of previous publications on that method and covers recent improvements for EAS elements. In particular, we describe three key issues of standard EAS elements and develop corresponding solutions.

The first issue concerns the low robustness of EAS elements in the Newton-Raphson scheme, which is characterized by the necessity for small load steps and many Newton iterations for the solution to converge. This behavior can be improved with the *mixed integration point* approach. That method was recently proposed for displacement-based elements and is extended to EAS elements in this work. The second issue deals with mesh distortion sensitivity, which is greatly reduced with a Petrov-Galerkin approach (unsymmetric stiffness matrix) in this dissertation. In fact, it can be shown that an unsymmetric stiffness matrix is a prerequisite to get elements with optimal performance for arbitrary element shapes (MacNeal's theorem). The newly proposed Petrov-Galerkin EAS ansatz is rather simple compared to many existing low-order Petrov-Galerkin methods and it enables to construct an element, which is exact in many bending problems regardless of element shape. Finally, we propose an alternative EAS framework for nonlinear kinematics based on enhancement of the spatial displacement gradient. It allows to eliminate the *spurious hourglass instabilities* of EAS elements, which have been under investigation for the last thirty years.

All in all, the newly proposed EAS element is robust, insensitive to mesh distortion and hourglassing-free. Thereby, it overcomes three of the most serious issues of present EAS elements and further improves their practical usability.

Keywords: enhanced assumed strain (EAS), mixed integration point, nonlinear preconditioner, Petrov-Galerkin, mesh distortion sensitivity, hourglassing-instabilities, stable finite elements, finite deformations

Kurzfassung

Bereits zu Beginn der 1990er wurde die *enhanced asssumed strain* (EAS) Methode vorgestellt, die eine der am weitesten verbreiteten gemischten Finite-Elemente-Methoden für die Festkörpermechanik ist. Die vorliegende kumulative Dissertation gibt einen ausführlichen Überblick über existierende Literatur zu dieser Methode und befasst sich außerdem mit einigen vor Kurzem eingeführten Verbesserungen für EAS-Elemente. Dabei werden drei bei EAS-Elementen auftretende Kernprobleme beschrieben und Lösungen dafür vorgestellt.

Das erste Problem betrifft die geringe Robustheit von EAS-Elementen im Newton-Raphson-Algorithmus, womit die hohe Anzahl an benötigten Lastschritten und Newton-Iterationen für die Konvergenz der Lösung bezeichnet wird. Dieses Verhalten kann mittels *mixed integration point* Ansatz verbessert werden. Diese Methode wurde vor Kurzem für Verschiebungselemente vorgestellt und wird in dieser Arbeit auf EAS-Elemente erweitert. Das zweite Problem betrifft die Empfindlichkeit bei Netzverzerrung, die in dieser Dissertation mit einem Petrov-Galerkin Ansatz (unsymmetrische Steifigkeitsmatrix) stark reduziert werden kann. Es existiert sogar ein Beweis, dass eine unsymmetrische Steifigkeitsmatrix eine notwendige Bedingung ist um optimale Ergebnisse unabhängig von der Elementform zu erhalten (MacNeals Theorem). Der neu vorgeschlagene Petrov-Galerkin EAS-Ansatz ist einfacher im Vergleich zu anderen Petrov-Galerkin Elementen mit niedriger Ansatzordnung und erlaubt die Konstruktion finiter Elemente, die unabhängig von der Elementform exakte Ergebnisse in vielen Biegeproblemen liefern. Abschließend wird ein alternatives EAS-Element für nichtlineare Probleme vorgestellt. Es basiert auf der Erweiterung des räumlichen Verschiebungsgradienten und ermöglicht es, die *künstlichen Hourglass-Instabilitäten*, die schon seit dreißig Jahren untersucht werden, zu beheben.

Zusammenfassend ist das neue EAS-Element robust, netzverzerrungsunempfindlich und frei von Hourglass-Instabilitäten. Damit werden einige der größten Nachteile von heutigen EAS-Elementen behoben und die praktische Anwendbarkeit kann dadurch weiter gesteigert werden.

Schlüsselwörter: enhanced assumed strain (EAS), mixed integration point, nichtlineare Vorkonditionierer, Petrov-Galerkin, Netzverzerrungsempfindlichkeit, Hourglass-Instabilitäten, stabile Finite-Elemente, große Deformationen

Acknowledgments

I had the pleasure of spending the last four and a half years as research associate and doctoral candidate at the Institute of Mechanics (IFM) of the Karlsruhe Institute of Technology (KIT). During that time I wrote this PhD thesis. My research was funded by the Deutsche Forschungsgemeinschaft (DFG, German Research Foundation), which is gratefully acknowledged. Furthermore, many people supported me and I want to take this opportunity to thank them.

First and foremost, I would like to express my deepest gratitude to my supervisor Prof. Peter Betsch for offering me a position as a research associate at the IFM and his support throughout my scientific venture. I have always been fascinated by his vast knowledge on mechanics and his ability to provide just the right reference or idea to make any project so much better. His passion for mechanics and numerics is contagious and inspired me to give my best. I would especially like to thank him for the freedom, encouragement and trust he gave me to pursue my own – sometimes unconventional – ideas.

My special thanks also go to Prof. Sven Klinkel (RWTH Aachen) for acting as second examiner of my thesis. Furthermore, I would like to extend my sincere thanks to Prof. Antonio Gil (Swansea University, UK) for supervising me during my four months in Swansea, which were funded by the Karlsruhe House of Young Scientists (KHYS). It was a pleasure to have countless fruitful discussions with him and Dr. Rogelio Ortigosa (UPCT Cartagena, Spain) and I enjoyed his never-ending enthusiasm and optimism despite the fact that the seemingly simple problem turned out to be a real challenge.

A crucial encounter for my scientific path took place at the congress on *Modern Finite Element Technologies (MFET)* in 2019: I met Simon Bieber (University of Stuttgart), who shares my love for EAS elements and became a good friend. It was invaluable to have another PhD student to discuss and exchange the latest research with and I am deeply indebted to him for suggesting a collaboration on a project, which now became part of this thesis. From the University of Stuttgart, I would also like to thank Prof. Manfred Bischoff for his helpful remarks on my work as well as the members of the scientific book club for reading about and discussing topics beyond the scope of our everyday research life.

Moreover, I would like to thank all my current and previous colleagues and friends at the IFM for their support in research and teaching. In particular I thank Dr. Marlon Franke for employing me as student and fostering my interest for numerical mechanics. Thanks go also to the members of the "IFM Running Club" (Alexander Janz, Tobias Laschütza, Paul Wasmer and Mark Schiebl) for the countless hard and competitive training sessions in the woods around the university, during which I had some of the best ideas for my research. Special thanks go to Tobias Laschütza and Timo Ströhle. Thank you for sharing the office with me, introducing me to (un-)pleasant music, distracting me when I needed it and most importantly becoming good friends. Big thanks go also to Paul Wasmer. Thank you for going through the challenges of teaching during the pandemic with me, your good-hearted spirit and your friendship. I would also like to particularly thank Moritz Hille for including me in his learning group and befriending me during our studies. Thank you for choosing me as supervisor for your master's thesis, which led to important insights for the present work. In general, I am deeply grateful to everyone at the institute for contributing to the great working environment and team spirit we had over the past years. Thank you for all the fun we had at the institute and at conferences, the many fruitful on and off topic discussions and the (sometimes too) long but always delightful evenings. I will forever happily remember the time we shared.

Finally, and most importantly, I would like to thank my family and friends. Words cannot describe my gratitude to my parents, who supported me throughout my life in every regard. Thank you for your advice and love, which made me who I am. Lastly, I am deeply grateful for all the support from my partner Marisa. Thank you not only for celebrating even my smallest breakthroughs, but especially for being at my side during the times when nothing seemed to work out. Without you I would have given up long ago. This thesis would not have been possible without you.

Karlsruhe, February 2023 Robin Pfefferkorn

Contents

1. Introduction

The rising level of complexity and optimization of modern engineering structures would not have been possible without powerful numerical simulation tools. Contemporary general-purpose and multi-physics numerical software allows to accurately predict the behavior of complex systems, which could previously only be studied with expensive and time-consuming experiments [154]. Two key ingredients facilitate these simulations: First, the ever-growing computational power, which allows to inexpensively solve more and more complex problems and enables higher resolution both in space and time. Second, the research effort put into developing and improving numerical methods, such as, e.g., the *finite volume method*, *finite difference method*, *boundary element method* and *finite element method* (FEM).

For solid mechanics, the FEM is arguably the most important method. Nowadays, there are hardly any civil or mechanical engineers not using finite element software to design their structures. The key idea of the FEM is to split the complex geometry of the problem into the name-giving elements of simple shape and compute an approximate solution using the weak form of a partial differential equation. When the FEM was established precisely is not easy to determine, as both mathematicians and engineers proposed parts crucial to the modern FEM in the early 20th century [162]. However, there seems to be a consensus that Clough (e.g., [33, 162]) was the first to use the term *finite element method*. One of the earliest works including most of the key ingredients of modern elements, namely numerical integration as well as a generalized isoparametric concept, was published by Irons [61] in 1966. Since then, there has been a plethora of publications, implementations and, of course, practical applications of the FEM.

A particular class of broadly applied finite elements are *low-order* elements with simple shapes and ansatz functions (e.g., triangles and quadrilaterals). The key advantages of such elements are their computational efficiency and simplicity in terms of mesh generation. Furthermore, low-order elements behave favorably when singularities arise, which are inevitable in real-world problems. This makes them especially suitable for nonlinear problems (see MacNeal [85] and Wriggers [154]). Unfortunately however, simple low-order elements suffer from severe *locking*, which denotes artificially high stiffness due to the numerical method. Locking of low-order displacement-based elements is in fact so severe, that they cannot be effectively applied without further remedies.

For *implicit analyses* (statics, slow/long-term dynamics) *mixed methods* are probably the most powerful tools to overcome locking. In contrast to the usual irreducible forms of differential equations, a mixed method incorporates additional unknowns which can be eliminated without loosing the well-definedness of the problem (see Zienkiewicz [162][1]). However, in numerical simulations it is possible to use these additional degrees of freedom to improve the numerical behavior of an element. One of the first[2] mixed finite elements was already proposed in 1965 by Herrmann [51], who introduced the pressure as independent variable in order to overcome *volumetric locking* (see also [27, 60, 135]). This type of locking is associated with (nearly) incompressible material models (e.g., rubbers and plastic deformations of metals) and is one of the two main types of locking in solid mechanic simulations. The other main type is *shear locking*, which arises in bending-dominated problems and cannot be cured with mixed methods based on pressure interpolations alone. Simultaneously curing both types of locking is, e.g., possible with mixed methods, that use the full stress and/or strain tensor as additional independent variable. Assumed stress elements, which use an independent stress field, are usually based on a *Hellinger-Reisner* variational principle and work extremely well in linear elasticity (e.g., [42, 110, 112]). Unfortunately, it is hardly possible to extend assumed stress elements to nonlinear problems as this requires an inverse stress-strain relation which scarcely exists (cf. [98, 145]). This problem can be overcome with an independent strain field. The resulting *strain-driven elements* harmonize well with the usual strain-driven constitutive equations making this type of element applicable in a wide range of examples [15]. The first class of such elements were the ad-hoc formulated incompatible mode models[3] first presented by Wilson et al. [149] in 1973 and later adapted by Taylor et al. [141] to pass the patch test. A generalization of incompatible mode models is the *enhanced assumed strain* (EAS) method presented by Simo and Rifai [134] in 1990. It is based on a *Hu-Washizu* variational framework and serves as a mathematically sound[4] justification of the incompatible mode models. Since the publication of the seminal work by Simo and Rifai [134], there has been a long history of publications on the EAS method (see, e.g., [47, 67, 69, 120, 131, 132] among many others). However, to keep the present introduction more concise, we refer to the introductions of the four publications reproduced in this dissertation for a comprehensive literature overview of the EAS method.

Due to the advantages described above, the EAS method has been implemented into several commercial FEM-codes and found wide application. However, there are still several open issues concerning EAS elements as shown by, e.g., Pfefferkorn and Betsch [107] and in Section 1.1. Thus, it is of high practical value to further improve EAS elements, which is the goal of the present cumulative dissertation.

[1] Other definitions of mixed methods are also possible. See, e.g., Boffi et al. [23] for a more mathematical view.

[2] See also Fraeijs de Veubeke [42] for a pioneering work on assumed stress elements.

[3] The additional field is a displacement field. However, in the weak form only its gradient appears, that is, as strain field.

[4] See Strang's and Taylor's famous dispute "*... two wrongs do make a right in California*" and "*... two rights make a right even California*" [134].

1.1. Motivation and methods

This motivation is based on the work of Pfefferkorn and Betsch [107], where three key issues of existing EAS elements are discussed. All three are addressed in-depth in this thesis and solutions/improvements are proposed for each of the problems.

Robustness

The first issue covered in this dissertation is the lack of robustness of EAS elements in the *Newton-Raphson* (NR) scheme. Poor robustness is characterized by the necessity of small load steps and a high number of NR-iterations to obtain convergence. Thus, in this sense, robustness also implies numerical efficiency, since less costly matrix factorizations are necessary. In this motivation we consider the simple clamped beam example [109] shown in Figure 1.1 for an illustration of the phenomenon. Comparing the required number of NR-iterations n_{NR} of the assumed stress element Q1/S5 [145] with the standard nonlinear EAS element Q1/E4 [131], reveals the poor robustness of the latter.

A simple remedy for that behavior is based on the *mixed integration point* (MIP) method (see Magisano et al. [88]), which is extended to EAS elements in [109] (see Chapter 3). The so-modified EAS element has substantially increased robustness similar to Q1/S5.

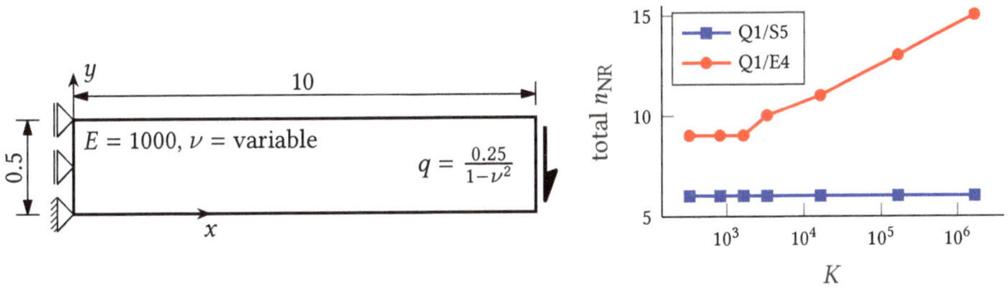

Figure 1.1.: Robustness in the Newton-Raphson scheme illustrated with a clamped beam example. Problem setup with geometry and boundary conditions (left). Required number of NR-iterations n_{NR} in dependence of the bulk modulus $K = E/(3 - 6\nu)$ with one load step for St. Venant-Kirchhoff material (right) [109].

Mesh distortion sensitivity

The second issue concerns sensitivity of EAS elements to mesh distortion. Again, a simple example (see, e.g., [106, 134]) shown in Figure 1.2 allows to analyze the problem. The linear elastic clamped beam is subjected to a pure bending moment and meshed with two increasingly distorted elements. Both the assumed stress element Q1/S5 [110] and the EAS element Q1/E4 [134] are able to give the analytic result $\delta = 1$ in case of a rectangular mesh. However, in case of distorted meshes the performance drops substantially. This behavior is in accordance with MacNeal's theorem [84] which states, that no element can simultaneously pass the patch test and be exact in higher order problems (e.g., bending) unless the stiffness matrix is unsymmetric. Constructing such elements is the key idea of the *unsymmetric FEM*, which was first proposed by Rajendran [117] for higher-order elements. Unfortunately, it is rather difficult to find suitable ansatz spaces for low-order finite elements.

A novel well-working unsymmetric low-order approach has recently been presented by Pfefferkorn and Betsch [106] (see Chapter 4), who proposed a Petrov-Galerkin EAS framework. That element is straightforward to construct and insensitive to mesh distortion without the serious drawbacks of earlier proposed unsymmetric low-order finite elements.

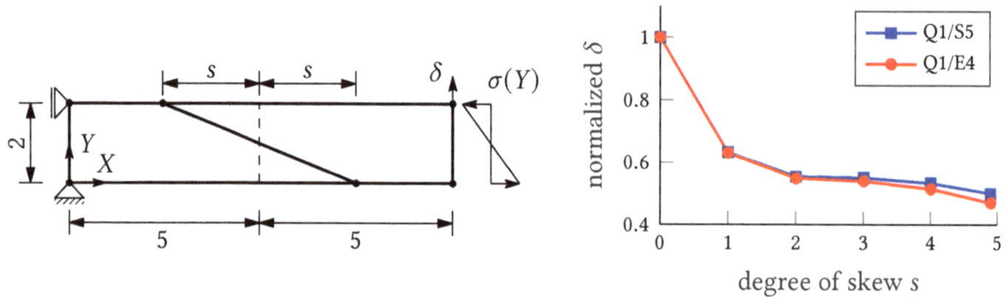

Figure 1.2.: Mesh distortion sensitivity. Setup of the mesh distortion test (left) and normalized displacement in dependence of distortion s (right) [106].

Hourglassing instabilities

The final issue considered in this work are spurious hourglassing instablities of EAS elements in nonlinear simulations, which were already mentioned in the conclusion of the first work on nonlinear EAS elements by Simo and Armero [131]. Wriggers and Reese [156] provide the first thorough explanation of that phenomenon for element Q1/E4 [131] for a hyperelastic block under compression. Fortunately, there is a relatively simple fix to overcome this problem in the absence of material instabilities. The solution is to use the *transpose* of the originally proposed *Wilson-modes* as ansatz for the enhanced field (see [47, 69], element Q1/E4T). However, a similar hourglassing phenomenon occurs in elasto-plastic simulations (and for other materials with instabilities) under tension as shown in Figure 1.3 for an elasto-plastic necking simulation (see, e.g., [8, 47, 105]). The test illustrates, that regardless of choice of ansatz functions, there is hourglassing for existing EAS elements. In fact, hourglassing in elasto-plastic simulations is not only a problem of EAS elements, but concerns other mixed elements as well, as shown in detail in the work of Hille et al. [52].

A solution to this problem is to enhance the spatial displacment gradient instead of the usual deformation gradient, which has recently been proposed by Pfefferkorn and Betsch [108] (see Chapter 5) for implicit problems and similarly by Schmied [125] for explicit problems.

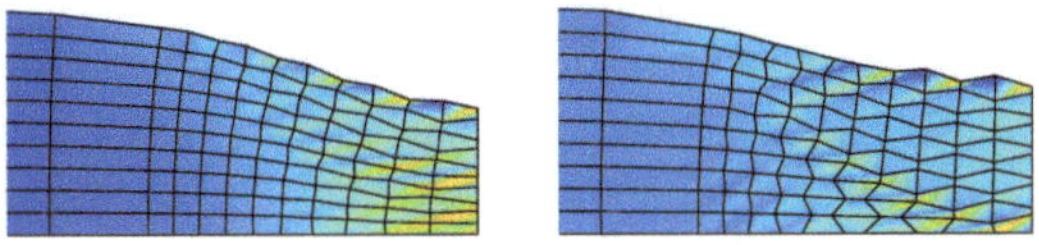

Figure 1.3.: Typical hourglassing patterns in a 2D elasto-plastic necking simulation. Results for Q1/E4 (left) and Q1/E4T (right) taken from [105].

1.2. Outline

The present cumulative dissertation is structured into six chapters. **Chapter 1** gives an introduction to the topic and describes the open issues solved with the novel approaches presented in the remainder of this work. Chapters 2 to 5 contain the cumulative part of the dissertation, each chapter reproducing one publication. Naturally, every work can be read independently. Therefore, repetitions and changes of notation are inevitable. In detail the chapters are structured as follows:

Chapter 2 reproduces [104]. This chapter gives a comprehensive overview of existing EAS finite elements, focusing on the various transformations and ansatz functions, that have been proposed for EAS elements (Section 2.2). Furthermore, we present novel approximations for both parts of the deformation gradient, which improve the element's performance[5]. Another main concern of Chapter 2 is the effect of various transformations and ansatz functions on the patch test, which is covered in Section 2.2.4. In particular, it is shown, that the modification of the compatible part of the deformation gradient for improved locking behavior (Simo et al. [132]) causes failure of the patch test, which is in contrast to a newly proposed correction. In general, Chapter 2 can be seen as summary of the status quo of EAS elements and as comprehensive introduction to the remainder of the work.

Chapter 3 reproduces [109]. This chapter addresses the first issue listed in Section 1.1. It thoroughly discusses the difference in robustness between assumed stress and EAS elements exemplified at a simple model problem in Section 3.3. Subsequently, the *mixed interpolation point* (MIP) method proposed by Magisano et al. [88] is extended to general material models in Section 3.4 and to EAS elements in Section 3.5. It is then shown with extensive numerical studies in Section 3.6, that the MIP approach can be used to improve robustness of EAS elements in many examples. Furthermore, Section 3.A introduces a novel inverse stress-strain relation for a Neo-Hookean material model.

Chapter 4 reproduces [106]. This chapter introduces a *Petrov-Galerkin EAS* framework for linear elasticity used to reduce mesh distortion sensitivity of EAS elements, that is, the second issue in Section 1.1. To that end we revisit MacNeal's theorem in Section 4.2

[5] Unfortunately, the novel transformation proposed for the enhanced deformation gradient leads to an instability in case of distorted meshes, which we discovered after publication of [104] (see [105, 108]).

and determine design conditions for optimal Petrov-Galerkin EAS elements in Section 4.3. The newly developed ansatz spaces to fulfill the design conditions are established in Section 4.4. Numerical investigations are presented in Section 4.5 and demonstrate the increased numerical performance obtained with the Petrov-Galerkin EAS ansatz. Besides the reduced mesh distortion sensitivity, further interesting features of the novel element are the exact numerical integration with standard Gauss quadrature and the equivalence to the standard EAS element [134] in case of regular meshes.

Chapter 5 reproduces [108]. This final chapter introduces a Petrov-Galerkin EAS framework for nonlinear solid mechanics overcoming all three major issues listed in Section 1.1. The key novelty of this chapter is the enhancement of the spatial displacement gradient introduced in Section 5.2 (see also Schmied [125]). Furthermore, the MIP approach developed in Chapter 3 and the Petrov-Galerkin ansatz from Chapter 4 are also employed (see Section 5.3) to get a robust and mesh distortion insensitive finite element. Properties and features of the newly proposed mixed element are listed in Section 5.4 before Section 5.5 covers extensive numerical studies, showing, that the element performs extremely well in many simulations. In particular, it is the only locking-free low-order element without artificial stabilization parameters, that is also hourglassing-free in all tests notoriously known to trigger instabilities for standard low-order mixed elements.

Chapter 6 concludes the present cumulative dissertation by summarizing the findings and giving an outlook to possible future research.

2. On transformations and shape functions for enhanced assumed strain elements

This chapter reproduces:*
Pfefferkorn R and Betsch P. "On Transformations and Shape Functions for Enhanced Assumed Strain Elements". In: *Int J Numer Meth Eng*. 120(2): 231–261, 2019. DOI: 10.1002/nme.6133

Abstract: We summarize several previously published geometrically nonlinear EAS elements and compare their behavior. Various transformations for the compatible and enhanced deformation gradient are examined. Their effect on the patch test is one main concern of the work, and it is shown numerically and with a novel analytic proof that the improved EAS element proposed by Simo et al. [132] does not fulfill the patch test. We propose a modification to overcome that drawback without loosing the favorable locking-free behavior of that element. Furthermore, a new transformation for the enhanced field is proposed and motivated in a curvilinear coordinate frame. It is shown in numerical tests that this novel approach outperforms all previously introduced transformations.

Keywords: enhanced assumed strain (EAS), finite deformations, hourglassing, mixed finite elements, patch test, transformations

2.1. Introduction

The computer simulation of large-scale solid mechanics problems requires robust, low-order and efficient general purpose finite elements. Such elements should be free of shear and volumetric locking, exhibit good coarse mesh accuracy and be insensitive to mesh distortion frequently induced by mesh generators (see Wriggers [154]). Additionally, no spurious (nonphysical) instabilities should arise from the element formulation. Unfortunately, low-order isoparametric displacement elements do not meet all of these criteria and show, e.g., severe locking in bending dominated problems as well as in the incompressible limit. Thus, there is a long history of various *mixed methods*, which incorporate extra fields such as strain or stress as primary variables in addition to the displacement.

A particular class of widely used mixed elements is based on the *enhanced assumed strain* (EAS) method, which was introduced for linear kinematics by Simo and Rifai [134] and for nonlinear kinematics by Simo and Armero [131] in the early 1990s. It is based on a *Hu-Washizu* type variational functional (see [148]) and its key idea is to introduce an enhanced strain field in addition to the compatible strain computed from the displacement field. This facilitates greatly reduced locking both in the nearly incompressible limit and in bending dominated problems[1]. The EAS method gives a mathematically solid foundation for the earlier introduced popular incompatible mode model by Wilson et al. [149], which was further improved by Taylor et al. [141] to pass the patch test. Note, that the element presented by Taylor et al. [141] is a special case of an EAS element. Moreover, all EAS-elements can be expressed as incompatible mode elements with the more complex scheme developed by Bischoff and Romero [22].
The EAS method provides a framework to construct finite elements with many desired properties. They can be constructed completely locking free, are relatively insensitive to mesh distortion, exhibit good coarse mesh accuracy and nonlinear material laws can easily be implemented due to the strain driven format. Thus, there is a plethora of publications on the method in various fields such as shell structures, modeling of failure and even diffusion problems (e.g., [3, 9, 17, 28, 71, 80]). In the present work, however, focus is put on two-dimensional and three-dimensional nonlinear solid mechanics (e.g., [8, 11, 12, 17, 36, 47, 63, 64, 66–70, 92, 95, 122, 124, 131, 132, 136, 138, 143, 147, 155] among others).
As mentioned above, the first geometrically nonlinear EAS element was proposed by Simo and Armero [131]. It is based on an enhancement of the deformation gradient and uses the nine (in 3D) *Wilson-modes* as shape functions for the enhanced deformation gradient. This element was then improved by Simo et al. [132] with three additional enhanced modes and a modification of the compatible deformation gradient to get a completely locking free element, which is important especially in elasto-plastic simulations as shown by Andelfinger et al. [4]. Unfortunately, the element presented in [132] does not fulfill the patch test (see [95, 155]). This issue will be thoroughly addressed in the present work. Another completely locking free element with a total of 21 enhanced modes, that passes the patch test was introduced by Andelfinger and Ramm [3]. A disadvantage of approaches using that many enhanced modes is the limited convergence radius of Newton's method due to the higher number of degrees of freedom from the higher order enhanced modes.
However, the probably most important drawback of the EAS method are nonphysical instabilities, which were already mentioned in the very first publication of Simo and Armero [131]. Interestingly, these instabilities arise only in the geometrically nonlinear case. In the linear case, the EAS method does not exhibit nonphysical instabilities, if three simple requirements are met (see [10, 120, 134]). However, the first nonlinear EAS element presented by Simo and Armero [131] exhibits an hourglassing instability under compression regardless of the material model used. This was first thoroughly covered and examined by Wriggers and Reese [156], Reese [121] for a Neo-Hookean material

[1] Note that there is a limit to improving bending behavior in distorted meshes as shown by MacNeal [83, 84].

and de Souza Neto et al. [37] for elasto-plasticity. These works showed that hourglassing occurs in a simple one element test case. More recently, Sussman and Bathe [138] showed, that these instabilities arise even in states of small strain, if the geometric aspect ratio of the elements is large. This renders EAS elements based on Wilson-modes almost unusable.
Thus, there have been many attempts to cure the instability, but none of which has proven to solve all problems so far. A first approach was presented by Korelc and Wriggers [69], whose element was adapted for objectivity by Glaser and Armero [47]. They proposed using the transpose of the Wilson-modes as shape functions for the enhanced deformation gradient and showed that the resulting element is free of spurious instabilities at least for the considered hyperelastic material. Unfortunately, for elasto-plastic and unstable elastic materials (see [47]) it still exhibits spurious hourglassing modes (see Armero [8]). Note, that the linearization of both, the modified element [47] and the standard EAS element based on Wilson-modes [131], yields the same geometrically linear EAS element [134].
Further attempts to remove the instability have been conducted in, e.g., [6, 8, 36, 47, 67, 123, 124, 147]. The elements of Glaser and Armero [47] and Korelc et al. [67] rely on the transpose Wilson-modes with further modifications, such as Taylor-expansion of the shape functions (see [68, 70]) in case of the element by Korelc et al. [67]. These elements are still not completely hourglassing free unless additional stabilization terms with arbitrary stabilization parameters are introduced. Such artificial stabilization techniques are also used, e.g., by Reese and Wriggers [124] and Reese [123], Wall et al. [147] and Areias et al. [6]. The latter work uses a penalty form of the usual variational framework for the EAS method. This penalty term has to be activated if hourglassing arises but then predates the benefits of the EAS method. Armero [8] introduced a combination of the EAS method by Glaser and Armero [47] with the mixed pressure element by Simo et al. [135]. The formulation further improves stability of the element by Glaser and Armero [47] but does not yield an unconditionally stable element (see [8]). Crisfield et al. [36] proposed enhancing the right stretch tensor instead of the deformation gradient which yields a very complex formulation due to the needed co-rotational finite element framework and seems to induce locking in some cases (see [154]). Another attempt made by Müller-Hoeppe et al. [95] splits the deformation into a homogeneous and inhomogeneous part. The instability is then removed by applying geometric nonlinear enhancement only to the homogeneous part and treating the enhancement of the inhomogeneous one with methods from linear theory. Apart from being nonphysical, major problems arise in the construction of material models for the method. A final possibility is to remove some critical enhancement modes as done, e.g., by Krischok and Linder [71], but this method removes the beneficial effects of those modes as well. Thus, it must be a case to case decision, if these modes are necessary to retain an efficient EAS element. All in all, there is so far no unconditionally stable EAS element without other major drawbacks.

The purpose of the present work is to summarize and compare shape functions and transformations previously used for EAS elements. One principal point is the modification of the compatible deformation gradient presented by Simo et al. [132], which leads to a violation of the patch test in generally distorted meshes (see [95, 155]). In the present work, we show that this failure can be cured with a simple modification which stems from the hourglass-stabilization introduced by Flanagan and Belytschko [40].We provide numerical and more importantly analytic proof, that the patch test is fulfilled by the newly developed approach. Furthermore, a new transformation between the reference and physical frame is proposed for the enhanced deformation gradient. This transformation is motivated with the help of a curvilinear coordinate system. Numerical investigations show the superior behavior of the novel approach compared to previously used formulas.

The present work is structured as follows. Section 2.2 briefly reviews the geometrically nonlinear EAS method and different approaches used within that framework. Focus is for that matter put on the finite element approximation of the compatible and enhanced deformation gradient in Section 2.2.2 and the patch test in Section 2.2.4. Details about the curvilinear framework to derive the novel transformation approach are given in Appendix 2.A. Additional information on computations needed to fix the violation of the patch test by the element of Simo et al. [132] is given in Appendix 2.B. Furthermore, Appendix 2.C covers the analytic proof, that the newly proposed improved version of the modification of Simo et al. [132] passes the patch test for arbitrarily distorted meshes. Numerical investigations to examine the properties of the elements introduced in Section 2.2 are given in Section 2.3, where standard benchmarks are used to assess the elements. Finally, a conclusion summarizes the work in Section 2.4.

2.2. Enhanced assumed strain method

This section briefly reviews the framework of the *EAS-method* for nonlinear kinematics, which was first presented by Simo and Armero [131]. Focus is put on the *finite-element* approximations of the deformation gradient. Various transformations for the compatible and incompatible field (see, e.g., [47, 71, 132, 134]) are summarized and partly motivated with the help of a *curvilinear* coordinate system. Furthermore, novel transformations are introduced for both parts of the deformation gradient.

2.2.1. Variational framework

The motion of a deformable body from its reference configuration $\mathcal{B}_0 \in \mathbb{R}^3$ to its current configuration $\mathcal{B} \in \mathbb{R}^3$ is described by a bijective deformation map $\boldsymbol{\varphi} : \mathcal{B}_0 \to \mathbb{R}^3$, which maps material points $\mathrm{X} \in \mathcal{B}_0$ onto spatial points $\mathrm{x} := \boldsymbol{\varphi}(\mathrm{X}) \in \mathcal{B}$. On a portion $\partial_\varphi \mathcal{B}_0$ of the boundary $\partial \mathcal{B}_0$ the displacement is prescribed by $\bar{\boldsymbol{\varphi}} : \partial_\varphi \mathcal{B}_0 \to \mathbb{R}^3$. Altogether the deformation map is given by

$$\boldsymbol{\varphi} \in \mathcal{U} = \left\{ \boldsymbol{\varphi} : \mathcal{B}_0 \to \mathbb{R}^3 \left| (\boldsymbol{\varphi})_i \in H_1, \ \det(\mathrm{D}\boldsymbol{\varphi}) > 0 \text{ and } \boldsymbol{\varphi}(\mathrm{X}) = \bar{\boldsymbol{\varphi}}(\mathrm{X}), \ \mathrm{X} \in \partial_\varphi \mathcal{B}_0 \right. \right\}. \quad (2.1)$$

The key idea of the geometrically nonlinear EAS method first presented by Simo and Armero [131] is to recast the deformation gradient $\mathbf{F}(\mathbf{X})$ at a material point $\mathbf{X}$ into the form

$$\mathbf{F}(\mathbf{X}) = \mathbf{F}_{\varphi}(\boldsymbol{\varphi}) + \tilde{\mathbf{F}}(\boldsymbol{\varphi}, \boldsymbol{\alpha}), \tag{2.2}$$

where[2] $\mathbf{F}_{\varphi} = \partial\boldsymbol{\varphi}/\partial\mathbf{X} = \mathrm{D}\boldsymbol{\varphi}(\mathbf{X})$ and $\tilde{\mathbf{F}}(\boldsymbol{\varphi}, \boldsymbol{\alpha})$ represent the compatible and enhanced part of the deformation gradient, respectively. Furthermore, the enhanced deformation gradient

$$\tilde{\mathbf{F}}(\boldsymbol{\varphi}, \boldsymbol{\alpha}) \in \tilde{\mathcal{F}} = \left\{\tilde{\mathbf{F}} : \mathcal{B}_0 \to \mathbb{R}^{3\times3} \left|(\tilde{\mathbf{F}})_{ij} \in L_2 \right.\right\} \tag{2.3}$$

is introduced as function of the deformation $\boldsymbol{\varphi} \in \mathcal{U}$ and n_{enh} enhanced parameters α_i, $i = 1, \ldots, n_{enh}$ arranged in vector $\boldsymbol{\alpha} \in \{\boldsymbol{\alpha} : \mathcal{B}_0 \to \mathbb{R}^{n_{enh}} | \alpha_i \in L_2\}$.

Remark 2.1. *Enhancing the deformation gradient* (2.2) *is not the only possible form of enhancement. Other kinematic measures can be enhanced analogously. Refer to, e.g., [17, 41, 66, 93, 147] (enhancement of the* Green-Lagrange-Strain*) and [36] (enhancement of the* right stretch tensor *combined with* co-rotational *approach).*

Remark 2.2. *Note, that the dependence of* $\tilde{\mathbf{F}}$ (2.3) *on* $\boldsymbol{\varphi}$ *and* $\boldsymbol{\alpha}$ *is a general case employed, e.g., in [47, 132] and can be simplified significantly, if the enhanced field can be derived from an incompatible displacement field (see Section 2.2.2.2 and [132]).*

Definitions (2.1) and (2.3) enable to give the *Hu-Washizu*-type *variational functional* of the EAS method for a hyperelastic material with *polyconvex strain-energy-function* $W(\mathbf{F})$ through (see [131])

$$\Pi_{\mathrm{EAS}}(\boldsymbol{\varphi}, \tilde{\mathbf{F}}, \mathbf{P}) = \int_{\mathcal{B}_0} W(\mathbf{F}) - \mathbf{P} : \tilde{\mathbf{F}} \, \mathrm{d}V + \Pi_{\mathrm{ext}}(\boldsymbol{\varphi}), \tag{2.4}$$

where $\Pi_{\mathrm{ext}}(\boldsymbol{\varphi})$ denotes the potential of external forces not further specified here. The third argument of Π_{EAS}, namely $\mathbf{P}$, is a stress-like variable and can be interpreted as *Lagrange multiplier* ensuring that condition $\tilde{\mathbf{F}} = \mathbf{0}$ holds in a continuous setting. It is defined by

$$\mathbf{P} \in \mathcal{P} = \left\{\mathbf{P} : \mathcal{B}_0 \to \mathbb{R}^{3\times3} \left|(\mathbf{P})_{ij} \in L_2 \right.\right\}. \tag{2.5}$$

The stationary conditions of (2.4) with respect to $\boldsymbol{\varphi} \in \mathcal{U}$, $\tilde{\mathbf{F}} \in \tilde{\mathcal{F}}$ and $\mathbf{P} \in \mathcal{P}$ are given in a next step. The corresponding admissible variations are $\delta\boldsymbol{\varphi} \in \mathcal{V}$, $\delta\tilde{\mathbf{F}} \in \tilde{\mathcal{F}}$ and $\delta\mathbf{P} \in \mathcal{P}$, where the set of admissible variations for the deformation $\boldsymbol{\varphi}$ has the form

$$\mathcal{V} = \left\{\boldsymbol{\varphi} : \mathcal{B}_0 \to \mathbb{R}^3 \left|(\boldsymbol{\varphi})_i \in H_1 \text{ and } \boldsymbol{\varphi}(\mathbf{X}) = \mathbf{0}, \mathbf{X} \in \partial_{\varphi}\mathcal{B}_0 \right.\right\}. \tag{2.6}$$

Since $\tilde{\mathbf{F}}$ was introduced as function of $\boldsymbol{\varphi}$ and $\boldsymbol{\alpha}$ in (2.3) it is straightforward to show that relation

$$\delta\tilde{\mathbf{F}} = \delta_{\boldsymbol{\varphi}}\tilde{\mathbf{F}} + \delta_{\boldsymbol{\alpha}}\tilde{\mathbf{F}} \tag{2.7}$$

[2] Note that here and subsequently arguments of functions are often omitted in favor of readability and notational simplicity.

holds. Therein $\delta_{(\bullet)}$ denotes the first variation of a functional with respect to $(\bullet)$. With this information at hand, the stationary conditions of (2.4) read

$$\delta_{\varphi}\Pi_{\text{EAS}} = \int_{\mathcal{B}_0} \hat{\mathbf{P}} : \delta_{\varphi}\mathbf{F} - \mathbf{P} : \delta_{\varphi}\tilde{\mathbf{F}} \, \mathrm{d}V + \delta_{\varphi}\Pi_{\text{ext}} = 0, \tag{2.8a}$$

$$\delta_{\alpha}\Pi_{\text{EAS}} = \int_{\mathcal{B}_0} \hat{\mathbf{P}} : \delta_{\alpha}\mathbf{F} - \mathbf{P} : \delta_{\alpha}\tilde{\mathbf{F}} \, \mathrm{d}V = 0, \tag{2.8b}$$

$$\delta_{\mathbf{P}}\Pi_{\text{EAS}} = -\int_{\mathcal{B}_0} \delta\mathbf{P} : \tilde{\mathbf{F}} \, \mathrm{d}V = 0, \tag{2.8c}$$

where $\hat{\mathbf{P}} = \partial_{\mathbf{F}} W(\mathbf{F})$ denotes the constitutive stress tensor.

Remark 2.3. *Note, that* (2.8c) *implies* $\tilde{\mathbf{F}} = \mathbf{0}$ *on body* $\mathcal{B}_0$ *in a continuous setting, which is ensured by* Lagrange multiplier $\mathbf{P}$. *Inserting* $\tilde{\mathbf{F}} = \mathbf{0}$ *into* (2.8a) *and* (2.8b) *reduces* (2.8) *to a pure displacement formulation showing consistency with continuum mechanics. However, this result holds only in a continuous, non-discrete setting. Using* (2.8) *as basis for a FE approximation can lead to improved numerical results.*

Remark 2.4. *Usually [131, 134],* (2.8) *is simplified by requiring* L_2*-orthogonality between the Lagrange-multipliers* $\mathbf{P}$ *and the enhanced deformation gradient* $\tilde{\mathbf{F}}$ *in a discrete setting. This makes the corresponding terms vanish in the equations above. For more details refer to Section 2.2.3.*

2.2.2. Finite element approximations

The *finite element method* (FEM) is one way to get an approximate solution of (2.8). In the present work only 4-node quadrilateral (2D) and 8-node brick (3D) elements Ω^e are considered for the discretization of body $\mathcal{B}$. The corresponding reference elements $\hat{\Omega} = [-1, 1]^{n_{dim}}$ are a bi-unit square and cube, respectively. Subsequently, focus is put on shape functions etc. for the three-dimensional case. They can however, easily be reduced to their two-dimensional equivalents.

2.2.2.1. Compatible deformation gradient

In this Section we introduce several procedures to compute the compatible deformation gradient. First, we briefly present the well-known isoparametric concept which is also the standard procedure used for EAS elements (see [131]). Second, we present a modification to that concept introduced by Simo et al. [132] which improves the locking behavior of the elements but leads to a violation of the patch test (see [6, 95, 155]). Finally, a new approach is presented which cures the failure of the patch test due to the modification of Simo et al. [132] but maintains its favorable locking behavior.

The standard procedure

Geometry and deformation are approximated[3] using the standard *isoparametric* concept (see [25, 154] for details). Thus,

$$\mathbf{X}^{\mathrm{h},e} = \sum_{I=1}^{8} N_I(\boldsymbol{\xi})\mathbf{X}_I^e, \qquad \boldsymbol{\varphi}^{\mathrm{h},e} = \sum_{I=1}^{8} N_I(\boldsymbol{\xi})\boldsymbol{\varphi}_I^e, \tag{2.9}$$

is defined for the approximations of geometry $\mathbf{X}$ and deformation $\boldsymbol{\varphi}$ within one finite element Ω^e. Therein, $\mathbf{X}_I^e$ and $\boldsymbol{\varphi}_I^e$ denote the nodal reference coordinates and deformation, respectively. Furthermore, standard *tri-linear* shape functions for the bi-unit cube with vertices (ξ_I, η_I, ζ_I) are employed. They are specified by

$$N_I(\boldsymbol{\xi}) = \frac{1}{8}(1 + \xi_I\xi)(1 + \eta_I\eta)(1 + \zeta_I\zeta), \quad I = 1, \ldots, 8, \tag{2.10}$$

and defined on the reference element $\hat{\Omega}$ in coordinate system $\boldsymbol{\xi} = [\xi, \eta, \zeta]^{\mathrm{T}}$. The variations $\delta\boldsymbol{\varphi}$ are approximated in the same way as $\boldsymbol{\varphi}$ in (2.9) yielding

$$\boldsymbol{\varphi}^h \in \mathcal{U}^h = \left\{ \boldsymbol{\varphi}^h \in \mathcal{U} \,\middle|\, \boldsymbol{\varphi}^h = \sum_{e=1}^{n_{el}} \boldsymbol{\varphi}^{\mathrm{h},e} \ \text{ and } \ \boldsymbol{\varphi}_I^e(\mathbf{X}_I^e) = \bar{\boldsymbol{\varphi}}^{\mathrm{h},e},\ \mathbf{X}_I^e \in \partial_\varphi \mathcal{B}_0^{\mathrm{h},e} \right\}, \tag{2.11}$$

$$\delta\boldsymbol{\varphi}^h \in \mathcal{V}^h = \left\{ \delta\boldsymbol{\varphi}^h \in \mathcal{V} \,\middle|\, \delta\boldsymbol{\varphi}^h = \sum_{e=1}^{n_{el}} \delta\boldsymbol{\varphi}^{\mathrm{h},e} \ \text{ and } \ \delta\boldsymbol{\varphi}_I^e(\mathbf{X}_I^e) = \mathbf{0},\ \mathbf{X}_I^e \in \partial_\varphi \mathcal{B}_0^{\mathrm{h},e} \right\}, \tag{2.12}$$

for the discrete ansatz spaces of the deformation and variations thereof (see [154]).

With this information at hand, transformations between the reference element and material configuration of a finite element Ω^e (see Figure 2.1) can be described by

$$\mathbf{J}^{\mathrm{h},e} = \frac{\partial \mathbf{X}^{\mathrm{h},e}}{\partial \boldsymbol{\xi}} = \sum_{I=1}^{8} \mathbf{X}_I^e \otimes \nabla_{\boldsymbol{\xi}} N_I, \tag{2.13a}$$

$$j^{\mathrm{h},e} = \det(\mathbf{J}^{\mathrm{h},e}), \tag{2.13b}$$

which denote the *Jacobian* matrix $\mathbf{J}^{\mathrm{h},e}$ of the transformation and its determinant $j^{\mathrm{h},e}$, respectively (see [134, 154]). Finally, the approximation of the compatible part $\mathbf{F}_\varphi$ of the deformation gradient (2.2) is given by

$$\mathbf{F}_\varphi^{\mathrm{h},e} = \mathrm{D}\boldsymbol{\varphi}^{\mathrm{h},e} = \frac{\partial \boldsymbol{\varphi}^{\mathrm{h},e}}{\partial \mathbf{X}^{\mathrm{h},e}} = \sum_{I=1}^{8} \boldsymbol{\varphi}_I^e \otimes \nabla_{\mathbf{X}} N_I, \tag{2.14}$$

where the derivatives of the shape functions with respect to $\mathbf{X}$ can be computed via well-known relation

$$\nabla_{\mathbf{X}} N_I = (\mathbf{J}^{\mathrm{h},e})^{-\mathrm{T}} \nabla_{\boldsymbol{\xi}} N_I. \tag{2.15}$$

[3] Approximations are denoted by superscript h.

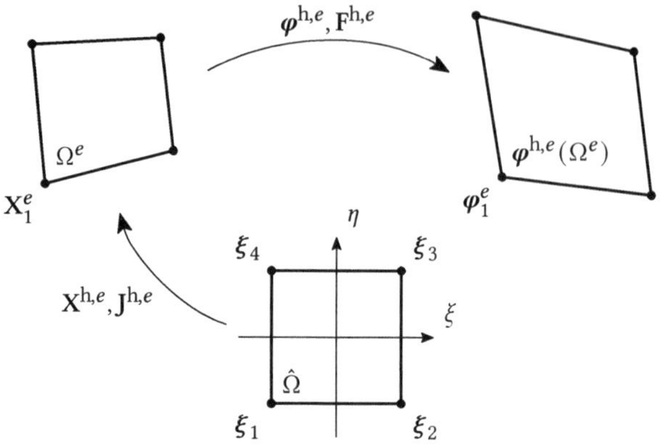

Figure 2.1.: Isoparametric map of a quadrilateral 2D finite element Ω^e.

Modification of the compatible deformation gradient by Simo et al.
Simo et al. [132] proposed a modified formula for the compatible deformation gradient to be used instead of (2.14). It allows to completely eliminate locking but unfortunately leads to violation of the patch test (see Sections 2.2.4 and 2.3.1, Appendix 2.C as well as References [6, 95, 155]). Nevertheless, the modification of the conforming deformation gradient is presented here to complete the overview of approximations for EAS elements. The idea proposed by Simo et al. [132] is to evaluate the gradients of the shape functions analogously to the enhanced field (see Section 2.2.2.2) and is based on a different representation of the shape functions used frequently for hourglass-stabilization (see, e.g., [16]). It ultimately leads to a modified gradient

$$\tilde{\nabla}_{\mathbf{X}} N_I = \nabla_0 N_I + \sum_{A=1}^{4} \frac{j_0}{j^{h,e}} \mathbf{J}_0^{-T} \nabla_\xi H_A \gamma_I^A, \qquad I = 1, \ldots, 8. \tag{2.16}$$

of the shape functions, which is then used in (2.14) instead of (2.15) to compute the conforming deformation gradient $\mathbf{F}_\varphi^{h,e}$. In the last equation,

$$\nabla_0 N_I := \nabla_{\mathbf{X}} N_I \Big|_{\xi=\mathbf{0}}, \tag{2.17}$$

$$\mathbf{J}_0 := \mathbf{J}^{h,e}\Big|_{\xi=\mathbf{0}} \quad \text{and} \quad j_0 := j^{h,e}\Big|_{\xi=\mathbf{0}}, \tag{2.18}$$

are the evaluations of (2.13) and (2.15) at the element center ($\xi = \mathbf{0}$). Moreover, $\nabla_\xi H_A$, $A = 1, \ldots, 4$ are gradients of the four *hourglass functions*

$$H_1 = \eta\zeta, \quad H_2 = \xi\zeta, \quad H_3 = \xi\eta, \quad H_4 = \xi\eta\zeta, \tag{2.19}$$

and γ_I^A are the components of the *gamma stabilization* vectors given by

$$\boldsymbol{\gamma}^A = \frac{1}{8}\left[\mathbf{h}^A - (\nabla_0 \mathbf{N})^{\mathrm{T}} \tilde{\mathbf{X}}^e \mathbf{h}^A\right], \qquad A = 1, \ldots, 4, \tag{2.20}$$

where

$$\nabla_0 \mathbf{N} = \begin{bmatrix} \nabla_0 N_1 & \cdots & \nabla_0 N_8 \end{bmatrix}, \tag{2.21}$$

$$\tilde{\mathbf{X}}^e = \begin{bmatrix} \mathbf{X}_1^e & \cdots & \mathbf{X}_8^e \end{bmatrix}, \tag{2.22}$$

$$\begin{aligned} \mathbf{h}^1 &= \begin{bmatrix} 1 & 1 & -1 & -1 & -1 & -1 & 1 & 1 \end{bmatrix}^{\mathrm{T}}, \\ \mathbf{h}^2 &= \begin{bmatrix} 1 & -1 & -1 & 1 & -1 & 1 & 1 & -1 \end{bmatrix}^{\mathrm{T}}, \\ \mathbf{h}^3 &= \begin{bmatrix} 1 & -1 & 1 & -1 & 1 & -1 & 1 & -1 \end{bmatrix}^{\mathrm{T}}, \\ \mathbf{h}^4 &= \begin{bmatrix} -1 & 1 & -1 & 1 & 1 & -1 & 1 & -1 \end{bmatrix}^{\mathrm{T}}. \end{aligned} \tag{2.23}$$

Refer to [132] for a more detailed description of the modification (2.16).

Remark 2.5. *Note, that the standard gradients of the tri-linear shape functions given by* (2.15) *are obtained in the context of this alternative notation by using*

$$\nabla_{\mathbf{X}} N_I = \nabla_0 N_I + \sum_{A=1}^{4} (\mathbf{J}^{h,e})^{-\mathrm{T}} \nabla_\xi H_A \gamma_I^A, \qquad I = 1, \ldots, 8. \tag{2.24}$$

instead of (2.16). *Thus, the modification presented by [132] is established by replacing* $(\mathbf{J}^{h,e})^{-\mathrm{T}}$ *in the original formulation* (2.24) *with* $\frac{j_0}{j^{h,e}} \mathbf{J}_0^{-\mathrm{T}}$. *We stress once more, that this modification leads to a violation of the patch test for distorted 3D meshes, which is shown numerically in Section 2.3.1 and analytically in Appendix 2.C.*

Remark 2.6. *In 2D, there is only one hourglass-function* $H_1 = \xi\eta$ *with corresponding vector* $\mathbf{h}_1 = \begin{bmatrix} 1 & -1 & 1 & -1 \end{bmatrix}^{\mathrm{T}}$ *and* gamma-stabilization *vector*

$$\gamma^1 = \frac{1}{4} \left[\mathbf{h}^1 - (\nabla_0 \mathbf{N})^{\mathrm{T}} \tilde{\mathbf{X}}^e \mathbf{h}^1 \right]. \tag{2.25}$$

Furthermore, gradient (2.15) *and* (2.16) *are always identical in 2D, which is also stated in [132]. This can be verified by comparing* (2.16) *and* (2.24), *which leaves*

$$j^{h,e} (\mathbf{J}^{h,e})^{-\mathrm{T}} \nabla_\xi (H_1) = j_0 \mathbf{J}_0^{-\mathrm{T}} \nabla_\xi (H_1) \tag{2.26}$$

to prove. The last equation can be established with lengthy but basic steps.

Improved version of the modification by Simo et al.

Finally, we present a new version to compute the compatible deformation gradient. It is based on the modification introduced by Simo et al. [132] (see above) and maintains its favorable locking behavior whilst curing the violation of the patch test. The only modification necessary is to replace $\nabla_0 N_I$ in (2.16) and (2.20) with

$$\bar{\nabla}_{\mathbf{X}} N_I = \frac{1}{V} \int_{\Omega^e} \nabla_{\mathbf{X}} N_I \, \mathrm{d}V, \qquad I = 1, \ldots, 8 \tag{2.27}$$

which is the average of the gradient of the shape functions within an element Ω^e with volume V. It can be evaluated efficiently with an analytic method developed by Flanagan and Belytschko [40] summarized in Appendix 2.B of the present work. The idea of using (2.27) stems from the works of Flanagan and Belytschko [40] and Belytschko et al. [16] and is well-known to cure the violation of the patch test in the context of hourglass-stabilization (see also [14, 128]). However, it has to the best of our knowledge[4] not been used for EAS elements so far. In Appendix 2.C we show analytically, that the proposed approach fulfills the patch test for arbitrarily distorted meshes in the present EAS framework.

The compatible deformation gradient is finally computed by using again (2.14) and replacing the standard gradient of shape functions $\nabla_{\mathbf{X}} N_I$ used there with

$$\hat{\nabla}_{\mathbf{X}} N_I = \bar{\nabla}_{\mathbf{X}} N_I + \sum_{A=1}^{4} \frac{j_0}{j^{\mathrm{h},e}} \mathbf{J}_0^{-T} \nabla_{\xi} H_A \hat{\gamma}_I^A, \qquad I = 1, \ldots, 8, \tag{2.28}$$

where the modified gamma-stabilization vectors are given by

$$\hat{\gamma}^A = \frac{1}{8} \left[\mathbf{h}^A - (\bar{\nabla}_{\mathbf{X}} \mathbf{N})^{\mathrm{T}} \tilde{\mathbf{X}}^e \mathbf{h}^A \right]. \tag{2.29}$$

Remark 2.7. *Without the modifications with the center evaluation of the Jacobian (see Remark 2.5),* (2.28) *is given by*

$$\hat{\nabla}_{\mathbf{X}} N_I = \bar{\nabla}_{\mathbf{X}} N_I + \sum_{A=1}^{4} (\mathbf{J}^{h,e})^{-\mathrm{T}} \nabla_{\xi} H_A \hat{\gamma}_I^A, \qquad I = 1, \ldots, 8. \tag{2.30}$$

Note, that this is in contrast to (2.24) *not equivalent to the standard gradient of shape functions* (2.15), *since a part of the shape functions is omitted as shown in Appendix 2.B. However, this has no negative implications on neither the accuracy of the method nor the completeness of the approximation spaces (see also Appendix 2.B and Belytschko and Bindemann [14]). On the contrary it provides the major advantage that it enables fulfilling the patch test for arbitrarily distorted meshes (see Appendix 2.C).*

2.2.2.2. Enhanced deformation gradient

After having described the approximation of the compatible part $\mathbf{F}_{\varphi}$ of the deformation gradient (2.2) in the last section, focus is now put on the approximation of the enhanced part $\tilde{\mathbf{F}}$. We list various requirements and review a general framework for a wide range of shape functions first presented by Simo et al. [132] (see also [47]).

The first requirement is *frame invariance*. By considering a superposed rigid body motion $\varphi^* = \mathbf{Q}\varphi + \mathbf{c}$ with the constant vector $\mathbf{c} \in \mathbb{R}^3$ and rotation matrix $\mathbf{Q} \in \mathrm{SO}(3)$ we note that

[4] During the review process we have been made aware of a similar but more complicated method proposed by Areias et al. [6].

the conforming deformation gradient $\mathbf{F}_{\varphi}$ transforms according to rule $\mathbf{F}_{\varphi}(\boldsymbol{\varphi}^*) = \mathbf{Q}\mathbf{F}_{\varphi}(\boldsymbol{\varphi})$. To get an overall *frame invariant* method, it is necessary that the complete deformation gradient (2.2) transforms accordingly. This ultimately yields requirement

$$\tilde{\mathbf{F}}(\boldsymbol{\varphi}^*, \boldsymbol{\alpha}) = \mathbf{Q}\tilde{\mathbf{F}}(\boldsymbol{\varphi}, \boldsymbol{\alpha}) \tag{2.31}$$

for the *incompatible* part of the deformation gradient. Usually, the incompatible part is approximated elementwise, to enable *static condensation* of the additional *degrees of freedom* and to get an efficient element. A possible and frequently used structure for the enhanced part of the deformation gradient is

$$\tilde{\mathbf{F}}^h \in \tilde{\mathcal{F}}^h = \left\{ \tilde{\mathbf{F}}^h \in \tilde{\mathcal{F}} \,\middle|\, \tilde{\mathbf{F}}^h = \sum_{e=1}^{n_{el}} \chi_e \tilde{\mathbf{F}}^h(\boldsymbol{\varphi}, \boldsymbol{\alpha}) = \sum_{e=1}^{n_{el}} \chi_e \mathbf{T}_0(\boldsymbol{\varphi}) \bar{\mathbf{F}}^e(\boldsymbol{\alpha}) \right\}, \tag{2.32}$$

which was first presented by Simo et al. [132] and is an elementwise approximation since

$$\chi_e = \begin{cases} 1, & \mathbf{X} \in \Omega^e \\ 0, & \text{else} \end{cases} . \tag{2.33}$$

The tensor $\mathbf{T}_0$ ensures frame invariance according to (2.31) if condition $\mathbf{T}_0(\boldsymbol{\varphi}^*) = \mathbf{Q}\mathbf{T}_0(\boldsymbol{\varphi})$ holds. Furthermore, $\mathbf{T}_0$ has to be elementwise constant in order to fulfill the patch test (see Section 2.2.4). A simple way to satisfy these conditions is setting

$$\mathbf{T}_0 = \mathbf{F}_0 := \mathrm{D}\boldsymbol{\varphi}^{\mathrm{h},e}\Big|_{\xi=0} = \mathbf{F}_{\varphi}^{\mathrm{h},e}\Big|_{\xi=0} \tag{2.34}$$

which corresponds to the evaluation of the conforming deformation gradient (2.14) at the element center [132]. However, other measures can be used. One novel version introduced in the present work employs the transposed inverse of $\mathbf{F}_0$ in the form

$$\mathbf{T}_0 = \mathbf{F}_0^{-\mathrm{T}}, \tag{2.35}$$

which is motivated in Appendix 2.A. It can easily be shown, that this choice of $\mathbf{T}_0$ fulfills the aforementioned requirements. Further possibilities used in previous publications are given in Remark 2.8.

Remark 2.8. *The center evaluation* (2.34) *is, e.g., used in [8, 47, 67, 132]. A possible alternative is the use of the average deformation gradient*

$$\mathbf{F}_{avg} = \frac{1}{V} \int_{\Omega^e} \mathbf{F}^{h,e} \, \mathrm{d}V \tag{2.36}$$

within an element with volume V as done in, e.g., [64, 71, 80, 95]. However, differences between using the average value and the evaluation at the element center are very small in usual problems without localization of strain (see Section 2.3.7). Other alternatives used in previous publications are the co-rotational approach using the rotation tensor $\mathbf{R}$ *at the element center emerging from the polar decomposition of the deformation gradient* $\mathbf{F}$ *(see Crisfield et al. [36] – enhancement is applied to the* right stretch tensor *instead of the deformation gradient in that work).*

The last part of the deformation gradient (2.2) that needs to be approximated, is the part $\bar{\mathbf{F}}^e$ introduced in (2.32). According to Glaser and Armero [47], a general form for arbitrary shape functions is given by

$$\bar{\mathbf{F}}^e = \sum_{I=1}^{n_{enh}} \mathbf{M}_I(\mathbf{X})\,\alpha_I, \tag{2.37}$$

where α_I are the n_{enh} enhanced parameters per element and $\mathbf{M}_I$ are the transformed shape functions. However, there are many possible transformations of the shape functions $\hat{\mathbf{M}}_I(\xi)$ defined on the reference element $\hat{\Omega}$ to the physical space. In general, they improve the bending behavior in initially distorted meshes. One well-working and widely used possibility (see, e.g., [8, 47, 63, 64, 69, 71]) is

$$\mathbf{M}_I(\mathbf{X}) = \frac{j_0}{j^{\mathrm{h},e}(\xi)}\mathbf{J}_0^{-\mathrm{T}}\hat{\mathbf{M}}_I(\xi)\,\mathbf{J}_0^{-1} \tag{2.38a}$$

where $j^{\mathrm{h},e}$, j_0 and $\mathbf{J}_0$ are given in (2.13b) and (2.18), respectively. The special structure with the elementwise constant quantities j_0 and $\mathbf{J}_0$ ensures that the patch test is fulfilled by construction of the enhanced field (see Section 2.2.4). Other possibilities, which are compared to (2.38a) in Section 2.3.7, are

$$\mathbf{M}_I = \frac{j_0}{j^{\mathrm{h},e}}\mathbf{J}_0\hat{\mathbf{M}}_I\mathbf{J}_0^{-1}, \tag{2.38b}$$

$$\mathbf{M}_I = \frac{j_0}{j^{\mathrm{h},e}}\mathbf{J}_0\hat{\mathbf{M}}_I\mathbf{J}_0^{\mathrm{T}}, \tag{2.38c}$$

$$\mathbf{M}_I = \frac{j_0}{j^{\mathrm{h},e}}\mathbf{J}_0^{-\mathrm{T}}\hat{\mathbf{M}}_I\mathbf{J}_0^{\mathrm{T}}, \tag{2.38d}$$

$$\mathbf{M}_I = \frac{j_0}{j^{\mathrm{h},e}}\mathbf{J}_0^{-\mathrm{T}}\hat{\mathbf{M}}_I, \tag{2.38e}$$

$$\mathbf{M}_I = \frac{j_0}{j^{\mathrm{h},e}}\mathbf{J}_0^{\mathrm{T}}\hat{\mathbf{M}}_I, \tag{2.38f}$$

$$\mathbf{M}_I = \frac{j_0}{j^{\mathrm{h},e}}\hat{\mathbf{M}}_I\mathbf{J}_0^{-1}, \tag{2.38g}$$

$$\mathbf{M}_I = \frac{j_0}{j^{\mathrm{h},e}}\hat{\mathbf{M}}_I\mathbf{J}_0^{\mathrm{T}}, \tag{2.38h}$$

$$\mathbf{M}_I = \frac{j_0}{j^{\mathrm{h},e}}\hat{\mathbf{M}}_I. \tag{2.38i}$$

All of the options listed above can furthermore be used with an elementwise average jacobian in analogy to (2.36) given by

$$\mathbf{J}_{\mathrm{avg}} = \frac{1}{V}\int_{\Omega^e} \mathbf{J}^{\mathrm{h},e}\,\mathrm{d}V \tag{2.39}$$

instead of the Jacobian $\mathbf{J}_0$ evaluated at the element center (2.18).

Remark 2.9. *Note, that* (2.38b) *is used in Simo et al. [132] and* (2.38c) *is similar to a* push-forward *of a* contravariant *tensor (used, e.g., by Kasper and Taylor [64]), whereas* (2.38a) *resembles* push-forward *of a* covariant *tensor. However, so far* (2.38a) *seems to be working best (see [47, 63] and Section 2.3.7). Refer to Appendix 2.A for a motivation based on curvilinear coordinates of some of the transformations listed above.*

The first set of shape functions $\mathbf{M}_I$ used in the present work for the enhanced field are the *Wilson-modes* presented by Wilson et al. [149]. They assume the form

$$\sum_{I=1}^{9} \hat{\mathbf{M}}_I \alpha_I = \begin{bmatrix} \xi\alpha_1 & \eta\alpha_2 & \zeta\alpha_3 \\ \xi\alpha_4 & \eta\alpha_5 & \zeta\alpha_6 \\ \xi\alpha_7 & \eta\alpha_8 & \zeta\alpha_9 \end{bmatrix} \tag{2.40}$$

in the current framework. Unfortunately, these shape functions lead to severe *hourglassing* in compression even if only simple hyperelastic materials are used (see, e.g., Section 2.3.4 and References [8, 37, 47, 69, 156]). Thus, Korelc and Wriggers [69] suggested using the transpose of (2.40) (see also [8, 47, 67]), as it removes *hourglassing* at least in compression and for hyperelastic materials. This is proven analytically in, e.g., [8, 47, 69].

Finally, the purely volumetric shape functions additionally introduced on top of the *Wilson-modes* by Simo et al. [132] are considered in the present work. They are given by

$$\sum_{I=10}^{12} \hat{\mathbf{M}}_I \alpha_I = (\eta\zeta\alpha_{10} + \xi\zeta\alpha_{11} + \xi\eta\alpha_{12})\,\mathbf{I}, \tag{2.41}$$

where $\mathbf{I}$ denotes the unity tensor. Note that there are no additional modes analogous to (2.41) in the two-dimensional case.

A very important requirement on the choice of shape functions for the enhancement is that the discrete ansatz-spaces for the deformation and the enhanced field may have no intersection. Namely, condition

$$\mathrm{Grad}[\mathcal{U}^h] \cap \mathcal{F}^h = \emptyset \tag{2.42}$$

has to hold for the discrete ansatz spaces $\mathcal{U}^h$ and $\mathcal{F}^h$ given in (2.11) and (2.32), respectively. This condition ensures stability of the method in linear analysis [120, 134] and is crucial in nonlinear simulations as well. However, other factors can still lead to instabilities in the nonlinear regime (see [8, 37, 47, 67, 69, 154, 156] among others).

Remark 2.10. *Note that* (2.40) *can alternatively be written as dyadic product of the form*

$$\sum_{I=1}^{9} \hat{\mathbf{M}}_I \alpha_I = \sum_{J=1}^{3} \boldsymbol{\beta}_J \otimes \hat{\mathbf{G}}_J(\boldsymbol{\xi}), \tag{2.43}$$

where $\boldsymbol{\beta}_1 = \begin{bmatrix} \alpha_1 & \alpha_4 & \alpha_7 \end{bmatrix}$ *and* $\hat{\mathbf{G}}_1 = \begin{bmatrix} \xi & 0 & 0 \end{bmatrix}$. *The other* $\boldsymbol{\beta}_J$ *and shape functions* $\hat{\mathbf{G}}_J$ *for* $J = 2, 3$ *are defined accordingly. Since* (2.43) *has the same structure as* (2.14) *(without transformations),* $\boldsymbol{\beta}_J$ *are referred to as incompatible displacement. Other types of shape functions cannot be put into this framework that simple, since they cannot be written in the same dyadic form. There is however always a possibility with more complex notation as shown by Bischoff and Romero [22]. If the shape functions for enhancement can be given in the simple form* (2.43) *it is possible to greatly simplify the discrete equations emerging from* (2.8). *Inserting* (2.43) *together with* (2.38a) *and* (2.34) *into* (2.32) *yields*

$$\tilde{\mathbf{F}}^{h,e} = \sum_{J=1}^{3} \frac{j_0}{j^{h,e}} \mathbf{F}_0 \mathbf{J}_0^{-\mathrm{T}} \left(\boldsymbol{\beta}_J \otimes \hat{\mathbf{G}}_J\right) \mathbf{J}_0^{-1}. \tag{2.44}$$

Defining $\boldsymbol{\gamma}_J = \mathbf{F}_0 \mathbf{J}_0^{-\mathrm{T}} \boldsymbol{\beta}_J$ *as new degrees of freedom finally recasts the last equation into the form*

$$\tilde{\mathbf{F}}^{h,e} = \sum_{J=1}^{3} \frac{j_0}{j^{h,e}} \boldsymbol{\gamma}_J \otimes \left(\mathbf{J}_0^{-\mathrm{T}} \hat{\mathbf{G}}_J\right) \tag{2.45}$$

which does not depend on $\boldsymbol{\varphi}$ *and has, apart from the multiplication with the determinants* j_0 *and* $j^{h,e}$*, the exact same structure as* (2.14) *(see [132]). This makes numerical implementation of elements employing shape functions of the form* (2.43) *especially easy, since variations and linearizations of* $\tilde{\mathbf{F}}^{h,e}$ *with respect to* $\boldsymbol{\varphi}$ *vanish. Note, that this scheme does not depend on* (2.38a) *and* (2.34) *but works with all other transformations listed in* (2.38) *and* (2.32).

2.2.2.3. Stress

The remaining field to be discretized for a full discrete form of (2.8) is the stress $\mathbf{P}$ defined in (2.5). Its approximation is done analogously to (2.32) and given by

$$\mathbf{P}^h \in \mathcal{P}^h = \left\{ \mathbf{P}^h \in \mathcal{P} \,\middle|\, \mathbf{P}^h = \sum_{e=1}^{n_{el}} \chi_e \mathbf{P}^{\mathrm{h},e}(\mathbf{X}) \right\}, \tag{2.46}$$

with the frame invariant interpolation functions $\mathbf{P}^{\mathrm{h},e}$, which have to include at least constant stress in order to fulfil the patch test (see Section 2.2.4 and Simo and Rifai [134]). The exact form of $\mathbf{P}^{\mathrm{h},e}$ is not needed, as the stress is eliminated via an orthogonality condition (refer to Section 2.2.3).

2.2.3. Orthogonality Condition

The stress field $\mathbf{P}^h$ is eliminated from the discrete version of (2.8) by setting $\mathcal{P}^h$ L_2-orthogonal to $\tilde{\mathcal{F}}^h$. Thus, the equation system (2.8) can be reduced to

$$\delta_{\varphi}\Pi^h_{\mathrm{mix}} = \int_{\mathcal{B}_0^h} \hat{\mathbf{P}}^h : \delta_{\varphi}\mathbf{F}^h \, \mathrm{d}V + \Pi^h_{\mathrm{ext}}(\delta\boldsymbol{\varphi}^h) = 0, \tag{2.47a}$$

$$\delta_{\alpha}\Pi^h_{\mathrm{mix}} = \int_{\mathcal{B}_0^h} \hat{\mathbf{P}}^h : \delta_{\alpha}\mathbf{F}^h \, \mathrm{d}V = 0, \tag{2.47b}$$

in a discrete setting, which is the basis for a finite element implementation. The relations covering orthogonality and elimination of the discrete stress presented here are based on the works of Simo et al. [131, 134].

2.2.4. Patch test condition

An important requirement for any finite element is that it fulfills the *patch test*, meaning that the response of an elastic solid subject to a constant state of strain is exactly reproduced for arbitrarily distorted patches of elements (see [142, 162]). For the classical EAS method this puts some restrictions on the shape functions $\hat{\mathbf{M}}_I$ introduced in (2.38) (see, e.g., [8, 131, 134]). Note, that the following statements are only valid if the compatible deformation gradient presented in Section 2.2.2.1 fulfills the patch test. Thus the compatible deformation gradient must be computed either in the standard way using (2.15) or with the modified version of approach (2.28) by Simo et al. The patch test is violated in distorted meshes, if the gradient of shape functions (2.16) presented by Simo et al. [132] is used. Refer to Appendix 2.C for more information on the patch test and the compatible field as well as an analytic proof, that the novel approach fulfills the patch test.

Two important conclusions can be drawn from the imposed constant strain $\mathbf{F}^{\mathrm{h},e} = \mathbf{F}_0$. First, they imply constant constitutive stress $\hat{\mathbf{P}} = \mathbf{P}_0 = \text{const.}$ for homogeneous materials, which are assumed in the present work. Second,

$$\tilde{\mathbf{F}}^{\mathrm{h},e} = \tilde{\mathbf{F}}^{e} = \mathbf{0} \qquad \Rightarrow \qquad \mathbf{F}^{\mathrm{h},e} = \mathbf{F}^{\mathrm{h},e}_{\varphi} = \mathrm{D}\boldsymbol{\varphi}^{\mathrm{h},e} = \mathbf{F}_0 \tag{2.48}$$

follows from condition (2.42) and the fact that the space of compatible deformation $\mathcal{U}^h$ (2.11) includes constant strain[5]. Note that $\delta\tilde{\mathbf{F}}^{\mathrm{h},e} \neq \mathbf{0}$ as the variations are arbitrary.

Ultimately, (2.47) has to be fulfilled exactly by the approximated fields to satisfy the patch test. Imposing conditions $\hat{\mathbf{P}} = \mathbf{P}_0 = \text{const.}$ and (2.48) on (2.47a) yields

$$\int_{\mathcal{B}_0^h} \mathbf{P}_0 : \mathrm{D}\delta\boldsymbol{\varphi}^h \;\mathrm{d}V + \Pi^h_{\mathrm{ext}}(\delta\boldsymbol{\varphi}^h) = 0 \tag{2.49}$$

for all admissible variations $\delta\boldsymbol{\varphi}^h$. This equation is fulfilled as it is equivalent to a pure displacement formulation, which satisfies the patch test if the approximation is chosen appropriately. Inserting the same relations into (2.47b) yields

$$\sum_{e=1}^{n_{el}} \mathbf{P}_0 : \int_{\Omega^e} \delta_{\alpha}\tilde{\mathbf{F}}^{\mathrm{h},e} \;\mathrm{d}V = 0. \tag{2.50}$$

With definitions (2.32), (2.37), (2.38) and by transforming the integral to the reference element we finally arrive at

$$\int_{\hat{\Omega}} \hat{\mathbf{M}}_I \;\mathrm{d}\hat{\Omega} = \mathbf{0} \tag{2.51}$$

[5] Since the isoparametric displacement element fulfills the patch test. This holds as well for modification (2.28) but not for (2.16).

which has to hold for all shape functions $\hat{\mathbf{M}}_I$ for the enhanced part of the deformation gradient. It is straightforward to show that this condition is fulfilled by shape functions (2.40) and (2.41). Note, that the form of (2.38) with the determinant of the Jacobian matrix (2.13b) in the denominator enables the simple form of (2.51).

2.3. Numerical investigations

This section covers standard numerical tests to evaluate the performance of the different transformations and shape functions presented in Section 2.2. Among the properties of interest are objectivity, sensitivity to mesh distortion, convergence behavior, stability of the elements and the patch test requirement. Furthermore, shear and volumetric locking are examined. All elements tested are based on 8-node brick elements in 3D and 4-node quadrilateral plane strain elements in 2D and use the standard 8- and 4-point quadrature rule if not stated otherwise.

The elements are identified as follows:

- **H1:** Three-dimensional isoparametric displacement formulation using the standard gradient of shape functions (2.15). The corresponding 2D element is denoted Q1.
- **HM1:** Isoparametric displacement formulation using the modified gradient of shape functions (2.16) introduced by Simo et al. [132].
- **H1/E9:** EAS element introduced by Simo and Armero [131] employing the Wilson-modes (2.40) for the enhanced field. Thus, there are nine additional enhanced parameters. The corresponding 2D element is denoted Q1/E4 and has 4 additional modes. If not stated otherwise, H1/E9 uses transformations (2.38a) and (2.34), which can be simplified for the given element (see Remark 2.10). Other transformations listed in (2.38) are denoted by suffixes of the form $J_0(x)$, where "x" is a letter from "a" to "i" corresponding to the transformations given in (2.38a) to (2.38i). Furthermore, J_0 is replaced by J_{avg} if the average jacobian (2.39) is used instead of the evaluation at the element center (2.18).
- **H1/E9T**: EAS element using the transpose of the Wilson-modes as introduced by Glaser and Armero [47].[6] The standard transformations used for this element are again (2.38a) and (2.34). Alternatives for transformation matrix $\mathbf{T}_0$ introduced in (2.32) are denoted by suffixes $F_0^{-\mathrm{T}}$ if the newly proposed option (2.35) is used and by "nonObj" for the non-objective choice $\mathbf{T}_0 = \mathbf{I}$. Other transformations regarding the shape functions of the enhanced modes, which are listed in (2.38), are denoted as described for element H1/E9.

[6] The transpose Wilson-modes were first suggested by Korelc and Wriggers [69] but the element presented there lacks frame invariance as shown in Glaser and Armero [47]. In the present framework, the non-objective element presented by Korelc and Wriggers [69] is denoted by H1/E9T-nonObj in 3D and Q1/E4T-nonObj in 2D.

- **HM1/E12:** Element with twelve enhanced modes as presented by Simo et al. [132]. It uses the Wilson-modes (2.40) and 3 additional modes (2.41) as well as the modified gradient (2.16) and a special 9-point quadrature rule (see, e.g., [67, 132]). The only difference of the corresponding 2D element QM1/E4 to Q1/E4 is the quadrature rule.
- **HM1/E12T:** Analogous to HM1/E12 but employing the transpose of the Wilson-modes (2.40) in addition to (2.41) as shape functions for the enhanced deformation gradient.
- **H1/E12** and **H1/E12T:** Equivalent to HM1/E12 and HM1/E12T but using the standard gradient (2.15) of the shape functions for the conforming part of the deformation gradient instead of the modified version (2.16). Suffix F_0^{-T} is added if the novel transformation (2.35) is applied for frame invariance of the enhanced modes.
- **HA1/E12** and **HA1/E12T:** Equivalent to HM1/E12 and HM1/E12T but using the improved version of the gradient of the shape functions given in (2.28), which fulfills the patch test in contrast to (2.16). Again, we use suffix F_0^{-T} to denote application of the novel transformation (2.35).

The material model used for all simulations in this work is a *Mooney-Rivlin* model with strain-energy function

$$W(\mathrm{C}) = a\,(I_1(\mathrm{C}) - 3) + b\,(I_2(\mathrm{C}) - 3) + \frac{c}{2}\left(\sqrt{I_3(\mathrm{C})} - 1\right)^2 - d \log\left(\sqrt{I_3(\mathrm{C})}\right), \qquad (2.52)$$

where $I_1(\mathrm{C})$, $I_2(\mathrm{C})$ and $I_3(\mathrm{C})$ denote the invariants of the right *Cauchy-Green-tensor* $\mathrm{C} = \mathrm{F}^{\mathrm{T}}\mathrm{F}$. The scalars a, b, c, with $a, b, c > 0$ for polyconvexity [24], are the three independent material parameters of the model and $d = 2a + 4b$ depends on the other parameters.

2.3.1. Patch test

The first test to assess the elements presented above is the standard patch test (see [86, 162]), where a deformation state is applied such that constant strain occurs. The required outcome is constant stress for homogeneous materials (see Section 2.2.4). It is an important condition, that any finite element has to fulfill (see [162]). For the EAS elements presented in the previous Section 2.2, it is shown in Section 2.2.4, that the patch test is fulfilled by design if condition (2.51) holds (which is the case for all shape functions considered here) and the compatible deformation gradient fulfills the patch test. This is the case if the standard gradient of shape functions (2.15) or modification (2.28) is used (see [16, 40]). However, employing (2.16) leads to a violation of the patch test. Corresponding analytic investigations are presented in Section 2.2.4 and Appendix 2.C. The numerical examples below are included to verify those results.

The test is performed on a unit cube $\mathcal{B}_0 = [0,1] \times [0,1] \times [0,1]$ in 3D and a unit square $\mathcal{B}_0 = [0,1] \times [0,1]$ in 2D, respectively. Dirichlet boundary conditions

$$u_i(X_i = 0, X_j, X_k) = 0 \qquad (2.53)$$

are applied on the lower surfaces ($X_i = 0$), implying that there is no deformation in the respective directions and yielding restraint free bearings. Constant strain is introduced by imposing an additional boundary condition

$$u_3(X_1, X_2, X_3 = 1) = u \tag{2.54}$$

on the upper surface $X_3 = 1$ (analogously in 2D with u_2 and $X_2 = 1$). Displacement u is increased in steps of $\Delta u = 0.05$ until the *Newton-Raphson* scheme finds no solution. The material parameters for the chosen Mooney-Rivlin material (2.52) are set to

$$a = 1.538 \cdot 10^5, \qquad b = 7.692 \cdot 10^4, \qquad c = 2.692 \cdot 10^5, \tag{2.55}$$

which correspond to $E = 10^6$ and $\nu = 0.3$ in linear theory. This leaves only the FE-mesh to complete the configuration of the patch test. A regular 3D mesh with $4 \times 4 \times 4$ elements is chosen in addition to an initially distorted[7] mesh (see Figure 2.2). The 2D meshes are chosen accordingly.

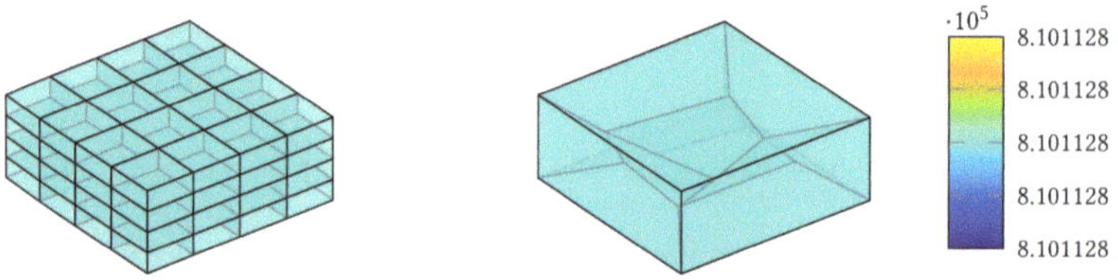

Figure 2.2.: Patch test. Three-dimensional regular (left) and distorted (right) mesh. Deformed configuration at $u = 0.55$ with von Mises stress distribution. Figures generated with H1 element.

As expected, all 2D elements and 3D elements using the standard isoparametric approach with standard gradient of the shape functions (2.15) or the improved version of the approach by Simo et al. (2.28) pass the patch test. This is in line with the investigations shown in Section 2.2.4, Remark 2.6 and Appendix 2.C. Unfortunately, 3D elements using the modified gradient (2.16) (HM1, HM1/E12 and HM1/E12T) only pass the patch test in case of the regular mesh, a fact first noted by Wriggers and Korelc [155] (see also Müller-Hoeppe et al. [95] and Areias et al. [6]).

The fulfillment of the patch test can be quantified by looking at the mean value and standard deviation of the *von Mises* stress of the modeled body. A typical element fulfilling the requirements shows for a displacement of $u = 0.4$ a mean value of $\sigma_{\text{mean}} = 4.978e5$ (exact value) and a small[8] standard deviation of approx. $\sigma_{\text{dev}} \approx 1e{-9}$ for any mesh. An element violating the patch test, such as, e.g., HM1, exhibits a slightly different mean value $\sigma_{\text{mean}} = 4.965e5$ and a high standard deviation $\sigma_{\text{dev}} \approx 1e4$ quantifying the violation of the test.

[7] The geometry of the distorted mesh is chosen according to [86].

[8] Due to round-off errors.

2.3.2. Objectivity

The goal of this test is to show that the elements are objective, meaning that their response is invariant to rigid body motions. The test is taken from [47] (see also, e.g., [67]) and its geometry is depicted in Figure 2.3. The beam-like structure has a length of $L = 1.0$ and its other dimensions are $h = b = 0.1$. Dirichlet boundaries are imposed on both ends of the beam with prescribed displacements

$$\mathbf{u}_1 = \mathbf{Q}(\theta_i)\mathbf{X}, \qquad \mathbf{u}_2 = \mathbf{Q}(\theta_i)\left(\mathbf{X} + u_i \mathbf{e}_y\right), \qquad i = 1, \ldots, n \tag{2.56}$$

assigned to the left and right end of the undeformed beam $\mathcal{B}_0$, respectively. They are applied during n steps in which the scalar displacement $u_i = u \cdot i/n$ is increased to the final magnitude $u = 2h$. Furthermore, rigid body motions are imposed on the beam through rotation matrix

$$\mathbf{Q}(\theta_i) = \begin{bmatrix} \cos(\theta_i) & -\sin(\theta_i) & 0 \\ \sin(\theta_i) & \cos(\theta_i) & 0 \\ 0 & 0 & 1 \end{bmatrix} \in SO(3), \qquad \theta_i = \theta \cdot i/n \tag{2.57}$$

where the final angles of rotation are chosen as $\theta \in \{0, 15, 30, 45, 60, 75, 90\}^\circ$ [47]. The number of load steps is $n = \theta/3 + 1$ for any given final angle θ. All final configurations of the continuum are shown in Figure 2.3 and were computed using a mesh with only six elements. Furthermore, the Mooney-Rivlin material (2.52) is used with parameters $a = 40$, $b = c = 10$, which correspond to $\lambda = 50$ and $\mu = 100$ or equivalently $E = 233.3$, $\nu = 0.1667$ in linear theory.

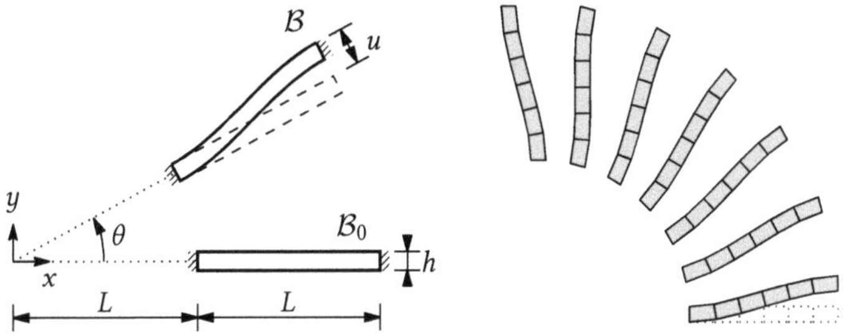

Figure 2.3.: Objectivity test. Geometry (left) and all final configurations (right). Reference configuration depicted with dotted lines. Plot on right side generated with 2D element Q1/E4.

Frame invariance is finally verified by examining the reaction forces at the bearings. This is done in a beam like manner by summarizing the nodal forces in the rotated local coordinate system into axial force N, shear force V and bending moment M. There may be no change in these values for an element to be objective, which is quantified by the standard deviation, of the reaction forces for different angles θ. All elements with the exception Q1/E4T-nonObj pass this test with standard deviations of the forces being below $1 \cdot 10^{-8}$. The clearly non-objective response of Q1/E4T-nonObj is depicted in Figure 2.4

and compared to the correct behavior of Q1/E4T. We stress that the newly proposed formula (2.35) for the transformation of the enhanced deformation gradient passes the present objectivity test.

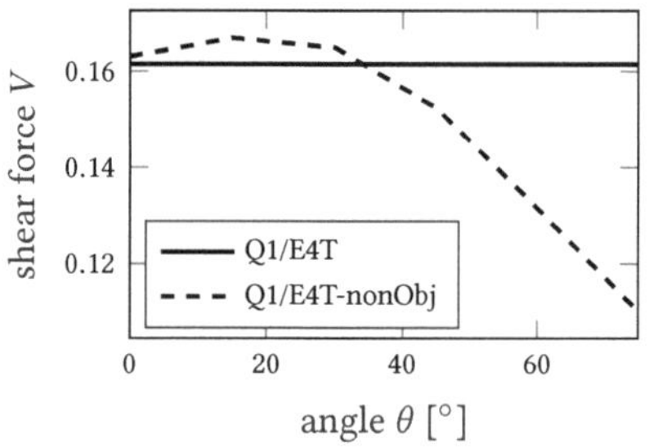

Figure 2.4.: Objectivity test. Demonstration of violation of objectivity by element Q1/E4T-nonObj. Shear force V at right bearing for different angles θ (see also [47]).

2.3.3. Linearized eigenvalue analysis

The next test is the linearized eigenvalue analysis first presented for EAS elements by Simo et al. [132] (see also [3, 64, 70, 115]). It is used to determine the locking behavior of elements in the incompressible limit, which is approximately enforced by setting

$$a = 0.35, \qquad b = 0.15, \qquad c = 1 \cdot 10^9 \tag{2.58}$$

for the Mooney-Rivilin material (2.52). This corresponds to a ratio $K/\mu = 1 \cdot 10^9$ for the bulk modulus K and shear modulus μ in linear theory. A single element, either distorted or regular (see Figure 2.5), is considered for this test and a spectral eigenvalue analysis is then conducted on its tangent matrix in the stress free reference configuration. This yields 24 modes for the 3D-elements and 8 modes for the 2D versions, which is equal to the respective number of displacement DOFs. Additional modes for the enhanced DOFs do not occur because static condensation (see, e.g., [131, 134]) eliminates these DOFs on element level.

The computed eigenvalues λ_i are summarized in three groups: rigid body modes with $\lambda_i = 0$, soft modes with finite λ_i and locking modes with[9] $\lambda_i \to \infty$. A completely locking free element should exhibit only one eigenvalue tending to infinity. The corresponding mode represents pure volumetric deformations (see [64, 132]). Furthermore, there have to be six rigid body modes with $\lambda_i = 0$ in 3D and only three in 2D, representing the number of possible independent motions of a rigid body. The right number of rigid body modes is recovered by all presented elements (see Table 2.1). All remaining modes should have finite eigenvalues and represent the deformation behavior of the elements.

[9] Infinity is not reproduced in the numeric analysis as incompressibility is only approximated by a high ratio of K/μ. This means that, in this benchmark, modes with "high" eigenvalues are considered as locking modes.

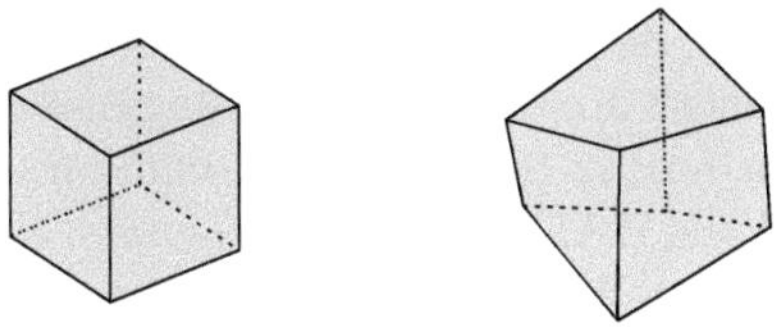

Figure 2.5.: Eigenvalue analysis. Regular (left) and distorted (right) element in 3D.

However, these ideal results are not obtained in the numeric analysis as most elements fail to reproduce the desired outcome with only one locking mode. The number of modes in each category is shown in Table 2.1 for some examined 3D elements.

Table 2.1.: Results of the spectral eigenvalue analysis of initial stiffness matrix, regular and distorted cube – selected elements.

element type	number of modes – regular			number of modes – distorted		
	rigid body[a]	soft modes[b]	locking modes[c]	rigid body[a]	soft modes[b]	locking modes[c]
H1	6	11	7	6	10	8
HM1	6	11	7	6	11	7
H1/E9	6	14	4	6	13	5
H1/E9T	6	14	4	6	13	5
H1/E9T-$F_0^{-\mathrm{T}}$	6	14	4	6	13	5
H1/E12	6	17	1	6	15	3
HM1/E12	6	17	1	6	17	1
HA1/E12	6	17	1	6	17	1
HA1/E12T-$F_0^{-\mathrm{T}}$	6	17	1	6	17	1

[a] number of spectral eigenvalues $|\lambda_i| \leq 1 \cdot 10^{-4}$
[b] number of spectral eigenvalues $1 \cdot 10^{-4} \leq \lambda_i \leq 1 \cdot 10^{2}$
[c] number of spectral eigenvalues $\lambda_i \geq 1 \cdot 10^{2}$ ("$\lambda_i \to \infty$")

Note, that the seven locking modes of H1 and HM1 for the regular cube show their severe locking, which gets even worse in distorted meshes for element H1 with an additional "high" mode. Enhancement with Wilson-modes (H1/E9) or the transpose thereof (H1/E9T, H1/E9T-$F_0^{-\mathrm{T}}$), which give identical results in this test[10], reduces locking but cannot completely eliminate it. This behavior is sometimes referred to as *mild locking* (see [132]) and has drawbacks especially if elasto-plastic materials are used (see Andelfinger et al. [4]). Using different transformations for the enhanced field listed in Section 2.2.2.2 does not change the number of locking modes in each category but slightly changes their values which renders more and less effective transformations. This is further elaborated upon in the Cook's membrane test in Section 2.3.7.

[10] Since the linearization of H1/E9 and H1/E9T yield the same linear element. In fact, both coincide with the original EAS element proposed by Simo and Rifai [134].

The only elements exhibiting the desired behavior with only one locking mode in both the regular and distorted mesh are the elements using one of the modified versions of the gradient of shape functions (2.16) or (2.28) in addition to twelve enhanced modes (HM1/E12(T), HA1/E12(T) and HA1/E12T$-F_0^{-T}$). However, element HM1/E12 as well as HM1/E12T, which use gradient (2.16), violate the patch test (see Section 2.2.4 and 2.3.1). If the modification of the compatible deformation gradient is omitted (e.g., H1/E12 and H1/E12T-F_0^{-T}), the element fulfills the patch test but shows two extra locking modes for the distorted cube. Only the novel approach based on (2.28) is able to maintain both - *completely* locking-free behavior for arbitrary meshes and fulfillment of the patch test.

In the 2D plane-strain case, the only elements subject to locking are Q1 and QM1. All other elements exhibit only one mode tending towards infinity and are thus completely locking free.

2.3.4. Stability analysis

The next test is the stability analysis introduced by Reese [121] to examine if elements are prone to hourglassing. This phenomenon has first been thoroughly covered for EAS elements by [121, 156] and is a major drawback of the original geometrically non-linear EAS element based on Wilson-modes presented in [131]. More recent results by Sussman and Bathe [138] show, that hourglassing is even possible for small strain if the geometric aspect ratio of the elements is large. This makes an hourglassing free element even more important. Note, that the stability test is often conducted in an analytic setting (see, e.g., [8, 47, 67, 121, 138, 154, 156]) in contrast to the numerical investigation presented in [37] and the present work. However, the same conclusions can be drawn.

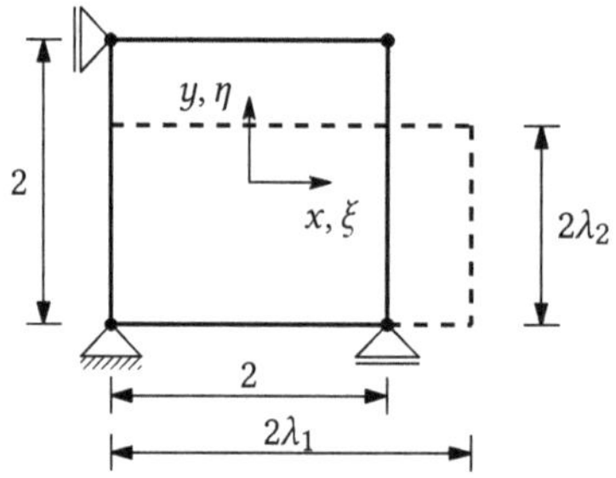

Figure 2.6.: Stability test (2D, extended in the present work for hexahedral elements). Geometry and constraints, deformed configuration depicted with dashed line.

The stability analysis is performed on a single element (see Figure 2.6) with exactly the same dimensions as the reference element presented in Section 2.2.2.1, making the transformations especially simple. The material parameters of model (2.52) are chosen to $a = 9$, $b = 1$ and $c = 99996$ representing a nearly incompressible material with $\mu = 20$ and $\lambda = 10^5$ (corresponding to a Poisson ratio of $\nu = 0.4999$ in the linear elastic case). Dirichlet boundaries are applied analogously to the patch test shown in Section 2.3.1, to gain a restraint free setting. The structure is loaded by imposing prescribed deformations on the top face ($y = 1$) and thereby determining the stretch λ_2. Despite being non-linear, an analytic solution for λ_1 can be found for the problem at hand. Thus, all displacements are determined and the exact deformation state can be imposed on the given element. Finally solving the eigenvalue problem

$$\mathbf{K}_{red}\mathbf{x}_i = \omega_i \mathbf{x}_i, \tag{2.59}$$

of the reduced tangential stiffness matrix $\mathbf{K}_{red}$ (excluding DOFs with imposed Dirichlet boundary conditions) for a given configuration yields the eigenvalues ω_i and corresponding eigenvectors $\mathbf{x}_i$. Instability points (singular points) are characterized by a zero eigenvalue. Furthermore, negative eigenvalues imply that the equilibrium is unstable (see [154]). Note, that there are also physical singular points for the given problem (see Glaser and Armero [47]) but they do not occur if the specimen is modeled with a single element (see [138] and Section 2.3.5).

All in all, the eigenvalues of $\mathbf{K}_{red}$ must fulfill $\omega_i > 0$, which implies that only the lowest eigenvalue ω_1 has to be examined to determine if any instabilities occur. The value of ω_1 is plotted against stretch λ_2 in Figure 2.7[11] for some tested EAS elements.

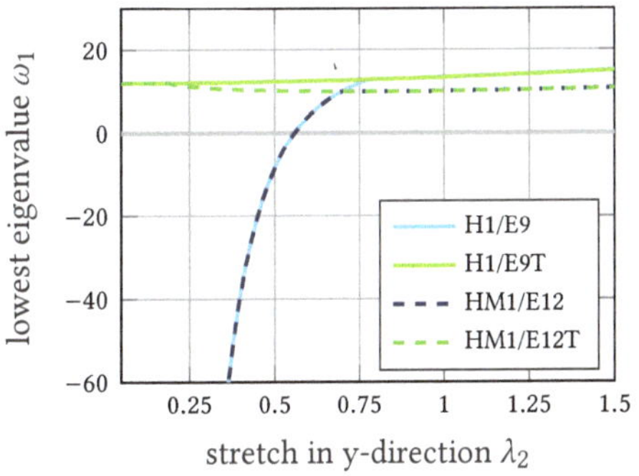

Figure 2.7.: Stability test. Selected 3D elements. Lowest eigenvalue ω_1 plotted against stretch λ_2.

Evaluating these plots shows that only elements employing the Wilson-modes for the enhanced field (2.40) exhibit negative eigenvalues. This is true whether the additional modes (2.41) are used or not. These instabilities occur under compression at $\lambda_2 \approx 0.61$ for the given geometry and material. Unfortunately, Sussman and Bathe [138] have shown that for elements with high aspect ratios instabilities already occur for small deformations with $\lambda_2 = 0.99$, which renders elements exhibiting negative eigenvalues almost unusable. Elements using the transpose of the Wilson-modes (H1/E9T, HM1/E12T) exhibit no negative eigenvalues and are therefore at least for the considered polyconvex hyperelastic materials stable. However, hourglassing still occurs for other kinds of materials, e.g., the frequently used elasto-plastic material by Simo [130]. This has for instance been reported

[11] Note, that states of compression are characterized by $\lambda_2 < 1$.

by Armero [8] for element Q1/E4T, which is unconditionally stable for elastic materials without material instabilities. The consideration of a elasto-plastic material model is however beyond the scope of the present work.

Note, that the type of transformation of the enhanced field and compatible field (see Sections 2.2.2.1 and 2.2.2.2) has no impact on the results listed above.

2.3.5. Large mesh stability test

This test is included to verify the results form the one-element stability analysis presented in Section 2.3.4 above with larger FE-meshes. This shows the elements' hourglassing behavior in combination with neighboring elements and also the ability to depict physical instabilities. The test presented here is mainly taken from the works of Glaser and Amero [47] and Korelc et al. [67]. It is performed with the same material properties as the stability test in Section 2.3.4 on a cube (see Figure 2.8) with an edge length of $a = 50$, which is discretized with twelve elements per side in a regular manner (see Figure 2.8).

Dirichlet boundary conditions $u_X = u_Y = u_Z = 0$ are applied on the corner node, one edge and lower surface as shown in Figure 2.8. The cube is loaded analogously to the one-element stability test by prescribed displacement $u_Z = u_i$ on the top surface. Again, an analytic (homogeneous) solution to the problem at hand can easily be found with a principal stretch based version of (2.52). This solution is imposed on the system by assigning these analytically obtained displacements to the nodes. Finally, an eigenvalue analysis (2.59) is performed on the stiffness matrix at that deformed state. Eigenvalues $\omega_i = 0$ characterize instability points which are determined approximately by gradually increasing the prescribed displacement u_i in steps of $\Delta u_i = 0.01a$. The procedure is repeated until either four stability points are passed or a displacement $u_i = 0.8a$ is reached. Determining and plotting the corresponding eigenvectors at states with $\omega_i \approx 0$ enables to decide if the instability is physical or unphysical (hourglassing).

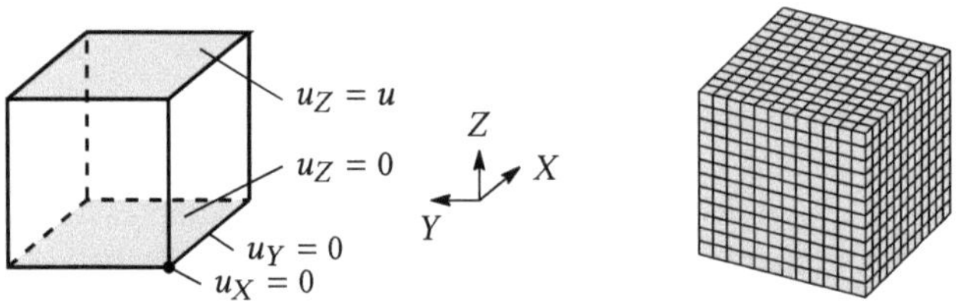

Figure 2.8.: Large mesh stability test. Geometry and mesh.

Four instability points are detected for all elements considered with the exception of the purely displacement-based elements, which are not able to reproduce that behavior. As suggested by the previous test in Section 2.3.4 the remaining elements can be classified into two main groups. The first one contains all elements based on Wilson-modes and show severe hourglassing. The second one contains the ones based on transpose Wilson-modes where only physical modes appear. Figure 2.9 shows the eigenmodes corresponding to the first four negative eigenvalues of H1/E9 and H1/E9T which are representative for their respective groups. All four nonphysical eigenvectors of H1/E9 depicted in Figure 2.9 occur at a state of compression of $\lambda_1 \approx 0.61$ ($u = 0.39$) which is in line with the results given in Section 2.3.4, where the instability was detected at exactly the same level. Physical instabilities of element H1/E9T first arise as the block is compressed to 55% of the initial height ($\lambda_1 \approx 55\%$). Using the additional enhanced modes (2.41) has only minor effects on the compression state at which the first instability occurs and does not change the shape of the computed eigenvectors. Note, that higher eigenmodes are affected more (e.g., the fourth instability occurs at $\lambda_1 = 0.42$ for H1/E9T and at $\lambda_1 = 0.51$ for HA1/E12TF_0^{-T}) but still the mode shape remains roughly the same. Furthermore, neither the type of transformations for the compatible deformation gradient in Section 2.2.2.1, nor the transformations listed in (2.38) have any effect on the results.

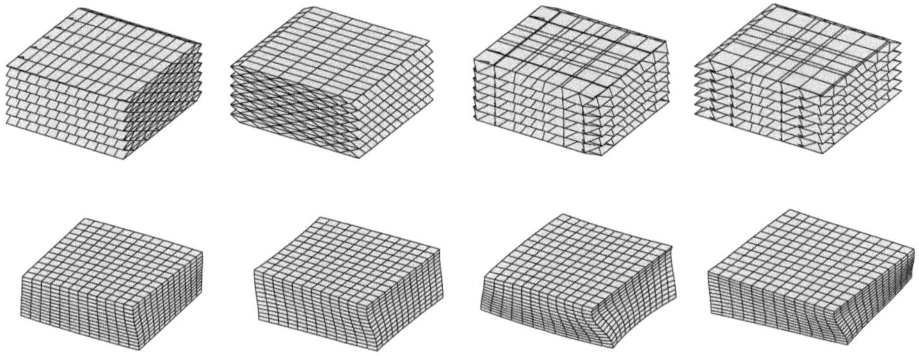

Figure 2.9.: Large mesh stability test. First four eigenmodes of H1/E9 (top) and H1/E9-T (bottom).

2.3.6. Mesh distortion

The next benchmark determines the element's behavior in distorted meshes. It is a standard benchmark included in, e.g., [2, 3, 6, 67, 115, 134] and usually performed on a cantilever-like structure, whose geometry and boundary conditions are shown in Figure 2.10. It is meshed with only two elements and mesh distortion is applied by moving the center nodes by s as indicated in Figure 2.10. Deformed and undeformed meshes for both the 2D and 3D case are shown in Figure 2.11. Note, that the nodes are moved unsymmetrically in the 3D case to get even more severely distorted elements. This is not the standard procedure but is applied here to get more insight into the distortion sensitivity of the elements. The material parameters of the Mooney-Rivlin model (2.52) are set to

$$a = 180, \qquad b = 120, \qquad c = 120, \tag{2.60}$$

which corresponds to the linear elastic parameters $E = 1500$ and $\nu = 0.25$. The test's setup is finished by applying a bending moment $M = 20$, assumed to be a dead load, on the right edge/face (see Figure 2.10). This results under the assumption of simple beam theory in a linear distribution of the stress

$$\sigma(Y) = -30 \cdot (Y - 1), \tag{2.61}$$

which is only dependent on material coordinates Y and based on the reference area (dead load). All in all, this yields nodal forces with a magnitude of $F = \pm M/h = \pm 10$ if there are only two nodes on the right edge as shown in Figure 2.10.

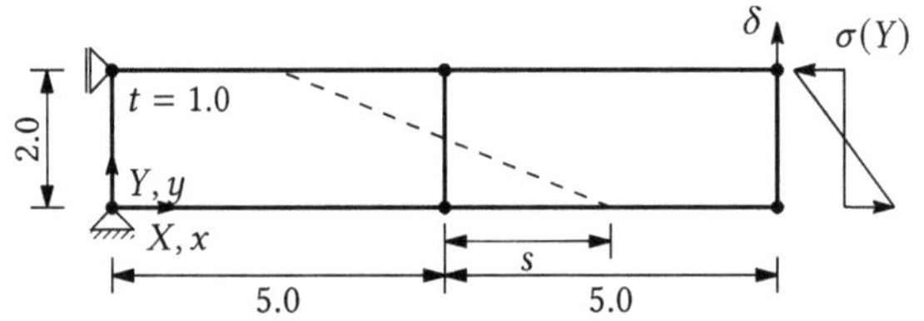

Figure 2.10.: Mesh distortion test. Geometry and boundary conditions.

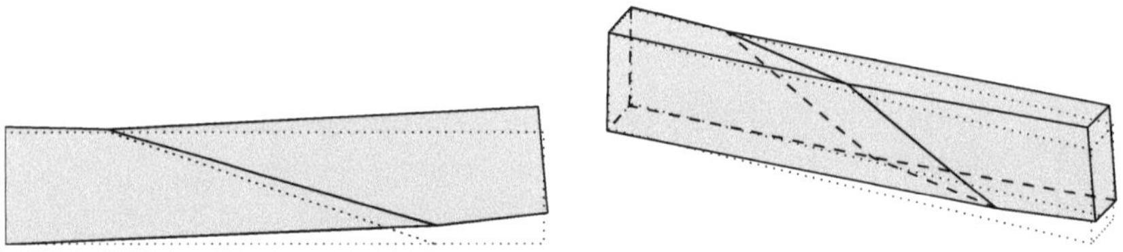

Figure 2.11.: Mesh distortion test. Deformed and undeformed configuration for $s = 3.0$ in 2D (left) 3D (right). Plots generated with standard EAS elements Q1/E4 and H1/E9.

The resulting normalized top edge displacements δ are presented in Figure 2.12, where they are plotted against the degree of skew s. Normalization of the displacements δ is conducted by comparing the computed deformations to the displacement δ_{ref} obtained from a simulation with a regular fine mesh with $40 \times 8 \times 4$ H1/E9 elements.

The plot in Figure 2.12a shows selected 2D elements. The first conclusion that can be drawn from this plot is that the displacement element Q1 exhibits severe locking. All other elements give almost the exact result for $s = 0$ with deviations from the converged result below 1%. There is a small difference between Q1/E4, Q1/E4T and QA1/E4T, QA1/E4T-F_0^{-T} which is due to the additional Gauss point of the latter elements. This coincides with the standard result for distorted isoparametric elements that the minimum number of Gauss points yields the best results (see Zienkiewicz et al. [162] Ch. 5.12). In the present 2D case 4 points would suffice but note that the case is different in 3D where the additional Gauss point is needed to integrate the volumetric enhanced modes (2.41) correctly (see [132]). Furthermore, it can be observed that the various approximations of the compatible deformation gradient (Section 2.2.2.1) and the novel transformation using $\mathbf{F}_0^{-T}$ (2.35) have only very little effect on the present results.

In the 3D case (Figure 2.12b), we observe again severe locking of the displacement formulation. Moreover, the effect of the distortion s is higher which is not surprising since the elements are more severely distorted due to the unsymmetric shift of nodes (see Figure 2.11). The elements with nine enhanced modes (e.g., H1/E9T) perform worse than elements with twelve modes in this case showing one advantage of the additional modes (2.41). The overall best element HM1/E12 is only slightly better than the element HA1/E12T-$F_0^{-\mathrm{T}}$ which in contrast to HM1/E12 fulfills the patch test and is stable for hyperelastic materials (see Sections 2.2.4, 2.3.1 and 2.3.4).

Finally, the effect of the transformations (2.38) could be examined. However, these results are qualitatively identical to the outcome of the Cook's membrane test in Section 2.3.7, which is why they are omitted here.

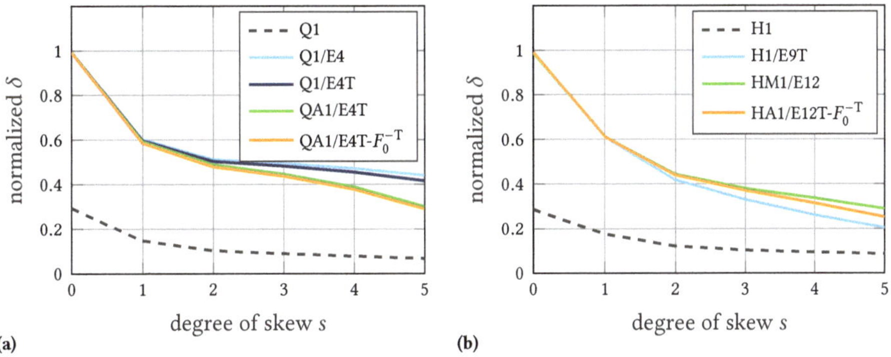

Figure 2.12.: Mesh distortion test. Normalized top edge displacement of selected 2D and 3D elements for different degree of skew s.

2.3.7. Cook's membrane

The final numerical test is the well-known Cook's membrane test (see, e.g., [8, 18, 19, 47, 63, 64, 126, 131, 134, 141]). It is designed to examine coarse mesh accuracy, convergence behaviour and shear locking in distorted meshes. Here, it is used to compare the performance of the various transformations and shape functions for the enhanced field presented in Section 2.2.2.

The general trapezoidal form and the clamped boundary condition on the left side of the Cook's membrane are depicted in Figure 2.13. Shear stress $\tau = 100$ is applied on the right edge, which is high enough to yield large deformations (see Figure 2.13). The parameters for Mooney-Rivilin material (2.52) are given by

$$a = 126, \qquad b = 252, \qquad c = 81661 \tag{2.62}$$

corresponding to $E = 2261$ and $\nu = 0.4955$ in linear theory [18]. Meshes used for the geometry described above always have two elements in direction of the thickness for 3D

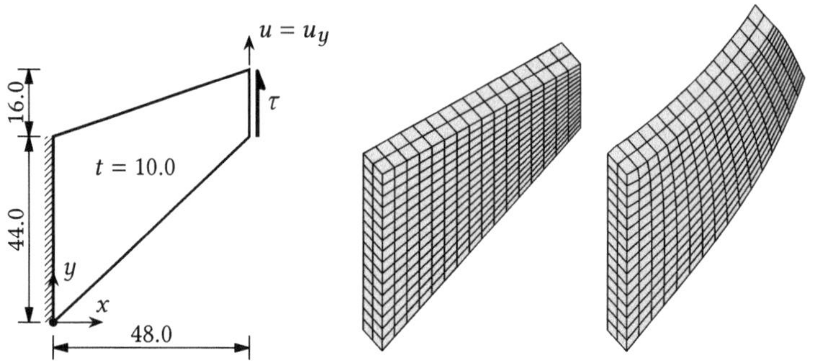

Figure 2.13.: Cook's membrane. Geometry and boundary conditions for the Cook's membrane test (left). Undeformed (middle) and deformed (right) configuration of Cook's membrane (3D). Plots generated with Q1/E9 element.

simulations and various numbers of elements per side n_{el}, namely $n_{el} \in \{2, 4, 8, 16\}$, to determine the convergence behavior with mesh refinement (h-convergence). The reference and deformed configuration for a 3D mesh with $n_{el} = 16$ is shown in Figure 2.13.

The evaluation of this test is conducted for several sets of elements, whose convergence behavior of the top corner displacement $u = u_y$ with mesh refinement is shown in Figure 2.14.

The results for the first set of elements, which includes various standard elements, are plotted in Figure 2.14a. Severe locking of the purely displacement-based element H1 is the first and most obvious result gained from the plot. Due to shear and volumetric locking it exhibits poor coarse mesh accuracy as well as slow convergence behavior, demonstrating impressively why enhanced elements were introduced in the first place. All other elements perform similar to H1/E9 with slightly better results for coarse meshes exhibited by H1/E12T, HM1/E12T and HA1/E12T due too the 3 additional enhanced modes. Interestingly, H1/E12T, HM1/E12T and HA1/E12T also perform better than H1/E12, HM1/E12 and HA1/E12 (not shown in the diagrams), respectively.

The second plot, Figure 2.14b, shows some elements based on the transpose of the Wilson-modes (2.40). It is included to compare the performance of the novel transformation, which uses the transpose inverse deformation gradient at the element center (2.35) for frame invariance, to standard EAS elements. Remarkably, element H1/E9T-F_0^{-T} with the newly proposed transformation performs 10.7% better for the coarsest mesh when compared to H1/E9T. There is even an improvement compared to HA1/E12T, even though H1/E9T-F_0^{-T} has three enhanced modes less. Note however, that in contrast to H1/E9T-F_0^{-T}, HA1/E12T is completely locking-free, which is advantageous in elasto-plastic tests (see [4]). The overall best performance in the Cook's membrane test of all elements tested in the present work can be accomplished with element HA1/E12T-F_0^{-T}, which uses the newly proposed improved modification of the compatible deformations (2.28), the novel transformation (2.35) and twelve enhanced modes.

The final plots, Figure 2.14c and 2.14d, compare the performance of various transformations listed in (2.38). Note, that there are many transformations which yield the same result for specific choices of shape functions. Only the unique transformations are included in the plots for both the EAS element based on Wilson-modes (Figure 2.14c) and transpose Wilson-modes (Figure 2.14d), respectively. All other transformations are duplicates of the ones shown in Figure 2.14, which can be proven with the information given in Remark 2.10 and also verified numerically. The most important result is that the standard transformation (2.38a) yields the best results for both types of shape functions. Furthermore, it is interesting that no transformation at all (2.38i) (elements H1/E9-$J_0(i)$ and H1/E9T-$J_0(i)$ i.e. original Wilson-modes and transpose thereof in combination with transformation (2.38i)) performs better than using an inappropriate transformation. For the examined case, there is virtually no difference between using the Jacobian at the element center (2.18) and the average Jacobian (2.39) (compare H1/E9-$J_0(a)$ and H1/E9-$J_{\mathrm{avg}}(a)$).

However, these results hold only for the special case of shape functions examined here. Other relations may be found for other shape functions than the Wilson-modes and the transpose thereof, which is why these tests should be repeated if novel modes for the enhanced field are proposed.

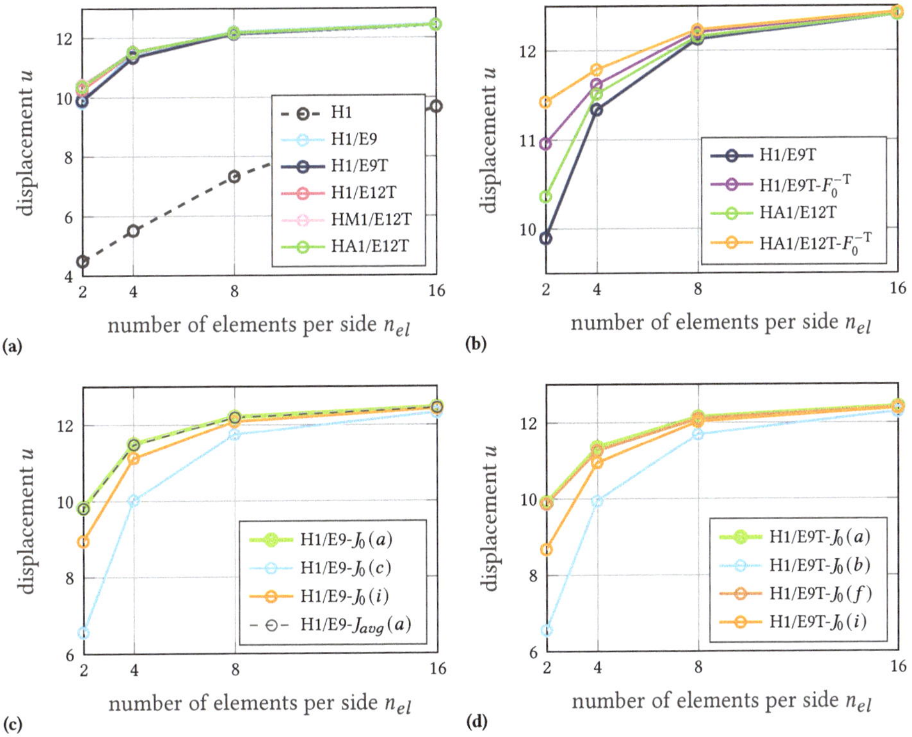

Figure 2.14.: Cook's membrane. Convergence of top corner displacement u with mesh refinement. Various sets of 3D elements.

2.4. Conclusion

First of all, we examined and compared shape functions previously used for the enhanced deformation gradient of the geometrically nonlinear EAS method and summarized their behavior in various tests. An important reproduced result is the by now well documented spurious hourglassing instability of the EAS element based on the original Wilson-modes (see Section 2.3.4 and [8, 37, 47, 69, 156]). This spurious behavior can be avoided for the polyconvex hyperelastic materials considered here by using the transpose of the Wilson-modes, which has been shown in previous works (see, e.g., [47, 69]). Moreover, we showed that the extra three enhanced modes introduced by Simo et al. [132] have little effect for the elastic problems considered here, even though they further reduce locking. For elasto-plastic problems however, this reduction of locking is crucial to compute correct limit loads (see Andelfinger et al. [4]).

Second, we thoroughly investigated transformations between the reference and physical domain for the compatible and incompatible deformation gradient. We showed numerically and with a novel analytic proof, that the modified evaluation of the compatible deformation gradient presented by Simo et al. [132] (e.g., element HM1/E12) does not fulfill the patch test in generally distorted meshes (see also [95, 155]). This makes the respective element unsuitable for general practical problems, where irregular meshes occur often due to mesh generators. This major drawback can be overcome with a novel approach we introduced in the present work. It relies on the evaluation of shape functions used for hourglass-stbilization introduced by Flanagan and Belytschko [40], which is employed here to modify the method of Simo et al. [132]. This novel approach has the advantage of maintaining the completely locking-free behavior of HM1/E12T for arbitrary meshes whilst curing the violation of the patch, which we showed with an analytic proof.
Finally, the transformation of the enhanced deformation gradient is also thoroughly covered in the present work and several novel transformations are introduced and compared. Most remarkably is that element H1/E9T-$F_0^{-\mathrm{T}}$, using an alternative transformation motivated with the help of a curvilinear coordinate system (see Appendix 2.A) to ensure frame invariance, yields superior coarse mesh accuracy compared to all previously introduced elements. Even HM1/E12T, which has 3 additional enhanced modes, performs worse than the element with the newly proposed transformation. Only elements HM1/E12T-$F_0^{-\mathrm{T}}$ and HA1/E12T-$F_0^{-\mathrm{T}}$, which use the novel transformation as well, perform slightly better.

All in all, the results of the present work show that EAS elements using the transpose Wilson-modes can be successfully employed to simulate elastic solid mechanics problems. We especially recommend using the newly proposed elements H1/E9T-$F_0^{-\mathrm{T}}$ and HA1/E12T-$F_0^{-\mathrm{T}}$ due to their excellent behavior in the problems considered here. However, the ultimate task of finding a generally applicable element requires elements also suitable for other geometries including shell structures and more importantly arbitrary materials. These issues are beyond the scope of the present work and should be investigated further.

Appendix to Chapter 2

2.A. Transformations of the deformation gradient

This appendix elaborates upon the transformation of the enhanced deformation gradient treated in Section 2.2.2.2 with the help of a curvilinear coordinate system, which is introduced by reinterpreting the isoparametric coordinates $\mathbf{X}^{\mathrm{h},e}$ given in (2.9) as functions of $\boldsymbol{\xi} = \begin{bmatrix} \xi & \eta & \zeta \end{bmatrix}^{\mathrm{T}} = \begin{bmatrix} \xi_1 & \xi_2 & \xi_3 \end{bmatrix}^{\mathrm{T}}$. The covariant basis vectors of the reference configuration (see Figure 2.15) of an element Ω^e are defined as

$$\mathbf{G}_i = \frac{\partial \mathbf{X}^{\mathrm{h},e}}{\partial \xi_i}, \qquad i = 1, 2, 3. \tag{2.63}$$

Additionally, a cartesian basis $\{\mathbf{e}_i\}$ is introduced in the space of the reference element $\hat{\Omega}$. Corresponding to $\{\mathbf{G}_i\}$, a dual contravariant basis $\{\mathbf{G}^i\}$ is defined by $\mathbf{G}_i \cdot \mathbf{G}^j = \delta_i^j$. Finally, a spatial covariant basis $\{\mathbf{g}_i\}$ is defined analogously to (2.63) by

$$\mathbf{g}_i = \frac{\partial \mathbf{x}^{\mathrm{h},e}}{\partial \xi_i}, \qquad i = 1, 2, 3, \tag{2.64}$$

and is derived from the spatial configuration $\mathbf{x}^{\mathrm{h},e} = \boldsymbol{\varphi}^{\mathrm{h},e}$. The tangential maps[12] between the configurations are given by

$$\mathbf{F}_{\varphi}^{\mathrm{h},e} = \sum_{i=1}^{3} \mathbf{g}_i \otimes \mathbf{G}^i, \qquad \mathbf{J}^{\mathrm{h},e} = \sum_{i=1}^{3} \mathbf{G}_i \otimes \mathbf{e}^i, \qquad \mathbf{j}^{\mathrm{h},e} = \sum_{i=1}^{3} \mathbf{g}_i \otimes \mathbf{e}^i, \tag{2.65}$$

which denote the approximations of the deformation gradient and jacobians (material and spatial). They satisfy relation $\mathbf{j}^{\mathrm{h},e} = \mathbf{F}_{\varphi}^{\mathrm{h},e}\mathbf{J}^{\mathrm{h},e}$. Further useful relations between the bases are, e.g.,

$$\mathbf{g}_i = \mathbf{j}^{\mathrm{h},e}\mathbf{e}_i, \qquad \mathbf{G}_i = \mathbf{J}^{\mathrm{h},e}\mathbf{e}_i, \tag{2.66a}$$

$$\mathbf{g}^i = (\mathbf{j}^{\mathrm{h},e})^{-\mathrm{T}}\mathbf{e}^i, \quad \mathbf{G}^i = (\mathbf{J}^{\mathrm{h},e})^{-\mathrm{T}}\mathbf{e}^i. \tag{2.66b}$$

All transformations and bases listed above are depicted in Figure 2.15.

In the next step, we assume that the enhanced deformation gradient $\tilde{\mathbf{F}}^{\mathrm{h},e}$ has the same mixed co-/contravariant and two-field structure as the compatible deformation gradient $\mathbf{F}_{\varphi}^{\mathrm{h},e}$, which has the advantage, that frame-invariance is automatically implied. Thus we define

$$\tilde{\mathbf{F}}^{\mathrm{h},e} = \tilde{F}^i_{.j}\mathbf{g}_i \otimes \mathbf{G}^j \tag{2.67}$$

[12] Tangential maps map infinitesimal line elements from one configuration to another.

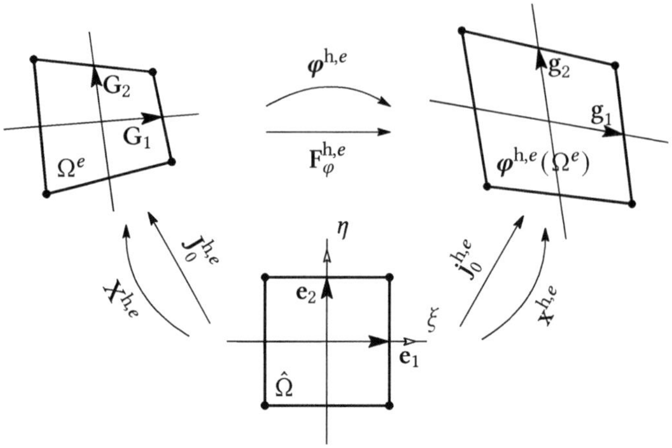

Figure 2.15.: Coordinate systems and transformations of an isoparametric quadrilateral 2D finite element Ω^e.

where the summation convention applies and the index with dot in $\tilde{F}^i_{.j}$ denotes the second index. Inserting from (2.66) into the last equation yields

$$\tilde{\mathbf{F}}^{\mathrm{h},e} = \tilde{F}^i_{.j}(\mathbf{j}^{\mathrm{h},e}\mathbf{e}_i) \otimes ((\mathbf{J}^{\mathrm{h},e})^{-\mathrm{T}}\mathbf{e}_j) = \mathbf{F}^{\mathrm{h},e}\mathbf{J}^{\mathrm{h},e}\left(\tilde{F}^i_{.j}\mathbf{e}_i \otimes \mathbf{e}_j\right)(\mathbf{J}^{\mathrm{h},e})^{-1}, \tag{2.68}$$

which describes the transformation of the enhanced field in the reference system $\tilde{\mathbb{F}} := \tilde{F}^i_{.j}\mathbf{e}_i \otimes \mathbf{e}_j$ into the physical space and was first introduced by Simo et al. [132]. In essence[13], (2.68) is the same as employing (2.38b) together with (2.34). Another possibility to define the enhanced field is the purely covariant version

$$\tilde{\mathbf{F}}^{\mathrm{h},e} = \tilde{F}_{ij}\mathbf{g}^i \otimes \mathbf{G}^j \tag{2.69}$$

which gives a transformation rule similar to combination of (2.38a) and (2.35) given by

$$\tilde{\mathbf{F}}^{\mathrm{h},e} = (\mathbf{F}^{\mathrm{h},e})^{-\mathrm{T}}(\mathbf{J}^{\mathrm{h},e})^{-\mathrm{T}}\tilde{\mathbb{F}}(\mathbf{J}^{\mathrm{h},e})^{-1}. \tag{2.70}$$

Note, that $\tilde{\mathbb{F}} = \tilde{F}^i_{.j}\mathbf{e}_i \otimes \mathbf{e}_j = \tilde{F}_{ij}\mathbf{e}_i \otimes \mathbf{e}_j$ since $\{\mathbf{e}_i\}$ is cartesian. Furthermore, (2.70) is also frame-invariant as described in Section 2.2.2.2. This consequent covariant approach has to the best knowledge of the authors never been used before. Instead, (2.38a) has often (see, e.g., [8, 47, 71]) been used together with (2.34) which can be viewed as mixture of the approaches (2.68) and (2.70) and yields a simpler implementation. Interestingly, approach (2.70) outperforms (2.68) in some numerical experiments, which is shown in Section 2.3.7. Which version is best, also depends on the type of shape functions chosen for the enhanced field.

[13] Without considering the center evaluation and scaling with determinants, which are necessary to fulfill the patch test (see Section 2.2.4).

2.B. Alternative representation of shape functions

This Appendix covers the computation of the average gradient of shape functions within an element defined in (2.27). Furthermore, we compare the various alternative representations of the gradient of shape functions used for the compatible deformation gradient given in Section 2.2.2.1. In particular, we adress a part of the shape functions that is neglected by the use of the average gradient. Note, that this neglected part has no implications on the patch test and does not lead to rank deficiencies of skewed elements (see Belytschko and Bindemann [14]). Using the average gradient of shape functions is even necessary to fulfill the patch test for generally distorted elements in (a) the context of hourglass-stabilization (c.f. Belytschko et al. [14, 16]) as well as for (b) the modification (2.28) of the gradient of shape functions used for the compatible deformation gradient in the present work (see Section 2.2.2.1 and Appendix 2.C).

2.B.1. Average gradient of shape functions

The analytic method to compute the average gradient of shape functions, which is outlined here, was developed by Flanagan and Belytschko [40] and can also be found in [16]. In a first step, the reference nodal coordinates $\mathbf{X}_I^e = \begin{bmatrix} X_I^e & Y_I^e & Z_I^e \end{bmatrix}^{\mathrm{T}}, I = 1, \ldots, 8$ are used to compute values b_I^X, b_I^Y and b_I^Z. They are given, e.g., for $I = 1$ by

$$
\begin{aligned}
b_1^X = \frac{1}{12}(\; & Y_2^e((Z_6^e - Z_3^e) - (Z_4^e - Z_5^e)) + Y_3^e(Z_2^e - Z_4^e) + \\
& Y_4^e((Z_3^e - Z_8^e) - (Z_5^e - Z_2^e)) + Y_6^e(Z_5^e - Z_2^e) + \\
& Y_5^e((Z_8^e - Z_6^e) - (Z_2^e - Z_4^e)) + Y_8^e(Z_4^e - Z_5^e) \quad),
\end{aligned} \tag{2.71a}
$$

$$
\begin{aligned}
b_1^Y = \frac{1}{12}(\; & Z_2^e((X_6^e - X_3^e) - (X_4^e - X_5^e)) + Z_3^e(X_2^e - X_4^e) + \\
& Z_4^e((X_3^e - X_8^e) - (X_5^e - X_2^e)) + Z_6^e(X_5^e - X_2^e) + \\
& Z_5^e((X_8^e - X_6^e) - (X_2^e - X_4^e)) + Z_8^e(X_4^e - X_5^e) \quad),
\end{aligned} \tag{2.71b}
$$

$$
\begin{aligned}
b_1^Z = \frac{1}{12}(\; & X_2^e((Y_6^e - Y_3^e) - (Y_4^e - Y_5^e)) + X_3^e(Y_2^e - Y_4^e) + \\
& X_4^e((Y_3^e - Y_8^e) - (Y_5^e - Y_2^e)) + X_6^e(Y_5^e - Y_2^e) + \\
& X_5^e((Y_8^e - Y_6^e) - (Y_2^e - Y_4^e)) + X_8^e(Y_4^e - Y_5^e) \quad).
\end{aligned} \tag{2.71c}
$$

Note, that only the coordinates X_I^e, Y_I^e, Z_I^e but not the node numbers I change in the right hand side of above equations. The values of b_I^X, b_I^Y and b_I^Z for $I \neq 1$ can be computed by permuting the nodes in (2.71) according to Table 2.2.

In a next step the exact volume of element Ω^e is computed from (2.71) via relation

$$
V = \sum_{I=1}^{8} b_I^X X_I^e. \tag{2.72}
$$

Table 2.2.: Permutations of nodes for b_I^X, b_I^Y and b_I^Z.

I	Permuted node numbers							
1	1	2	3	4	5	6	7	8
2	2	3	4	1	6	7	8	5
3	3	4	1	2	7	8	5	6
4	4	1	2	3	8	5	6	7
5	5	8	7	6	1	4	3	2
6	6	5	8	7	2	1	4	3
7	7	6	5	8	3	2	1	4
8	8	7	6	5	4	3	2	1

Using b_I^Y and Y_I^e or b_I^Z and Z_I^e in the equation above would yield the same result. With (2.71) and (2.72) at hand, the final relation to compute (2.27) is

$$\bar{\nabla}_{\mathrm{X}} N_I = \frac{1}{V}\begin{bmatrix} b_I^X & b_I^Y & b_I^Z \end{bmatrix}^{\mathrm{T}} \qquad I = 1,\ldots,8. \tag{2.73}$$

2.B.2. Comparison of alternative shape function representations

In the sequel we examine two representations of the tri-linear shape functions to quantify the inaccuracy introduced by the use of the average gradient of shape functions. We start by noting that the standard form of the tri-linear shape functions in the isoparametric space given in (2.10) can alternatively be expressed in vector form (e.g., Wriggers [154])

$$\mathrm{N} = \frac{1}{8}\mathrm{a}^0 + \frac{1}{8}\sum_{I=1}^{3}\mathrm{a}^I \xi_I + \frac{1}{8}\sum_{A=1}^{4}\mathbf{h}^A H_A(\boldsymbol{\xi}). \tag{2.74}$$

Therein, $H_A(\boldsymbol{\xi})$ and $\mathbf{h}^A$, $A = 1,\ldots,4$, are defined in (2.19) and (2.23), respectively. Moreover, the vectors a^I, $I = 0,\ldots,3$, are given by

$$\begin{aligned}
\mathrm{a}^0 &= \begin{bmatrix} 1 & 1 & 1 & 1 & 1 & 1 & 1 & 1 \end{bmatrix}^{\mathrm{T}},\\
\mathrm{a}^1 &= \begin{bmatrix} -1 & 1 & 1 & -1 & -1 & 1 & 1 & -1 \end{bmatrix}^{\mathrm{T}},\\
\mathrm{a}^2 &= \begin{bmatrix} -1 & -1 & 1 & 1 & -1 & -1 & 1 & 1 \end{bmatrix}^{\mathrm{T}},\\
\mathrm{a}^3 &= \begin{bmatrix} -1 & -1 & -1 & -1 & 1 & 1 & 1 & 1 \end{bmatrix}^{\mathrm{T}}.
\end{aligned} \tag{2.75}$$

2.B.2.1. Standard Taylor series representation of the shape functions

The Taylor series of the shape functions in the physical frame at the element center $\mathrm{X}_0 = \sum_{I=1}^{8} N_I\big|_{\boldsymbol{\xi}=\mathbf{0}}\mathrm{X}_I^e = \frac{1}{8}\tilde{\mathrm{X}}^e\mathrm{a}^0$ (with $\tilde{\mathrm{X}}^e$ defined in (2.22)) up to the linear part is given by

$$\mathrm{N}(\mathrm{X}^e) = \mathrm{N}\Big|_{\mathrm{X}^e=\mathrm{X}_0} + \frac{\partial \mathrm{N}}{\partial \mathrm{X}^e}\Big|_{\mathrm{X}^e=\mathrm{X}_0}(\mathrm{X}^e - \mathrm{X}_0) + \mathrm{N}_\gamma(\mathrm{X}^e). \tag{2.76}$$

Therein, the first two terms on the right-hand side furnish the linear part N_{lin}, while N_γ denotes the residual term. Taking a closer look at the linear part reveals

$$
\begin{aligned}
N_{lin} &= N\Big|_{X^e=X_0} + \frac{\partial N}{\partial X^e}\Big|_{X^e=X_0}(X^e - X_0) = \frac{1}{8}a^0 + (\nabla_0 N)^T\left(\tilde{X}^e N - \frac{1}{8}\tilde{X}^e a^0\right) \\
&= \frac{1}{8}a^0 + (\nabla_{\xi=0}N)^T J_0^{-1}\tilde{X}^e \frac{1}{8}\begin{bmatrix} a^1 & a^2 & a^3\end{bmatrix}\xi + \frac{1}{8}(\nabla_0 N)^T\tilde{X}^e\sum_{A=1}^{4} h^A H_A(\xi) \\
&= \frac{1}{8}a^0 + \sum_{I=1}^{3} a^I\xi_I + \frac{1}{8}(\nabla_0 N)^T\tilde{X}^e\sum_{A=1}^{4} h^A H_A(\xi),
\end{aligned}
\tag{2.77}
$$

where use has been made of relation

$$
\nabla_{\xi=0}N = \frac{1}{8}\begin{bmatrix} a_1 & a_2 & a_3\end{bmatrix}^T,
\tag{2.78}
$$

as well as $\partial N/\partial X|_{X^e=X_0} = (\nabla_{\xi=0}N)^T J_0^{-1}$ and $J_0 = \tilde{X}^e(\nabla_{\xi=0}N)^T$. The latter two relations can be established from (2.17), (2.13a) and (2.18). In view of (2.76) and (2.77) the residual part N_γ finally assumes the simple form

$$
N_\gamma = N - N_{lin} = \sum_{A=1}^{4}\frac{1}{8}\left[h^A - (\nabla_0 N)^T\tilde{X}^e h^A\right]H_A(\xi) = \sum_{A=1}^{4}\gamma^A H_A(\xi)
\tag{2.79}
$$

where the gamma-stabilization vectors (2.20) occur.

2.B.2.2. Using the average gradient of shape functions

The gradient of shape functions at the element center in (2.76) can be replaced with the average gradient of shape functions $\bar{\nabla}_X N$ defined in (2.27) and thorougly described in Section 2.B.1. This yields

$$
N(X^e) = N(X^e = X_0) + \left(\bar{\nabla}_X N\right)^T(X^e - X_0) + \hat{N}_\gamma(X^e).
\tag{2.80}
$$

Note, that this is not a real Taylor-series since the mean-value theorem does not hold for vector-valued functions and thus in general no X^* exists at which $\bar{\nabla}_X N = \nabla_X N(X^*)$ holds. The implications of this are described below.

The linear part N_{lin} cannot be simplified to the same extent as in the previous Section since the jacobians cannot be canceled out as in (2.77). In fact, with the same steps as before, the remaining residual part is

$$
\begin{aligned}
\hat{N}_\gamma &= \sum_{J=1}^{3}\frac{1}{8}\left[a^J - (\bar{\nabla}_X N)^T\tilde{X}^e a^J\right]\xi_J + \sum_{A=1}^{4}\frac{1}{8}\left[h^A - (\bar{\nabla}_X N)^T\tilde{X}^e h^A\right]H_A(\xi) \\
&= \sum_{J=1}^{3}\hat{\beta}^J\xi_J + \sum_{A=1}^{4}\hat{\gamma}^A H_A(\xi),
\end{aligned}
\tag{2.81}
$$

where we introduced

$$\hat{\beta}^J = \frac{1}{8}\left[\mathbf{a}^J - (\bar{\nabla}_X\mathbf{N})^T\tilde{\mathbf{X}}^e\mathbf{a}^J\right] \quad \text{and} \quad \hat{\gamma}^A = \frac{1}{8}\left[\mathbf{h}^A - (\bar{\nabla}_X\mathbf{N})^T\tilde{\mathbf{X}}^e\mathbf{h}^A\right]. \tag{2.82}$$

In contrast to (2.79) there are two residual parts in (2.81). The second part containing $\hat{\gamma}^A$ is almost the same as (2.79) with only the gradient of shape functions exchanged in the definition of the gamma-stabilization vector. The first one is linear and has a similar structure. As stated in Belytschko and Bindemann [14] this part is usually neglected for hourglass stabilization techniques based on the average gradient of shape functions and is also neglected in the present work (see (2.28)). However, this has no major impact on the accuracy and does not lead to insufficiently rich approximation spaces [14].

Remark 2.11. *In the case of a parallelepiped it is obvious from definitions* (2.27) *and* (2.17) *that* $\bar{\nabla}_X N_I \equiv \nabla_0 N_I$ *holds since the Jacobian is constant for such a geometry. Consequently, the vectors* $\hat{\beta}^J$ *and therefore the surplus part vanish, which can be shown with similar steps as in* (2.77).

2.C. Patch test and the compatible deformation gradient $\mathbf{F}_\varphi$

This appendix elaborates upon the patch test criterion for the compatible deformation gradient. In particular, we prove that the element proposed by Simo et al. [132] does not fulfill the patch test for arbitrary meshes while the newly proposed method (2.28) always fulfills it. We start by considering the homogeneous deformation

$$\boldsymbol{\varphi} = \mathbf{F}_0\mathbf{X} + \mathbf{c} \tag{2.83}$$

with constant tensor $\mathbf{F}_0 \in \mathbb{R}^{3\times3}$ and vector $\mathbf{c} \in \mathbb{R}^3$, which implies a constant deformation gradient[14]

$$\mathbf{F} = \frac{\partial\boldsymbol{\varphi}}{\partial\mathbf{X}} = \mathbf{F}_0. \tag{2.84}$$

The patch test criterion requires, that states of constant strain are reproduced exactly by the numerical method. Thus, under a state of homogeneous deformation with nodal deformations $\boldsymbol{\varphi}_I^e = \mathbf{F}_0\mathbf{X}_I^e + \mathbf{c}$ two conditions need to be met:

- First, $\mathbf{F}^{\mathrm{h},e} \equiv \mathbf{F}_0$ has to hold and
- second, it is necessary that the finite element residual is equivalent to the analytic solution, which is quite difficult to prove from scratch. However, the condition can easily be checked by comparison with the isoparametric element, since that element always fulfills the patch test.

[14] In general: $\mathbf{F}$ = const. $\Leftrightarrow$ $\boldsymbol{\varphi}$ = homogeneous (see Gonzales and Stuart [48]).

In the following we examine these requirements for all three versions of the compatible deformation gradient considered in the present work. Namely, we study the isoparametric concept with standard gradient of shape functions (2.15) (which is well-known to fulfill the patch test), the modification proposed by Simo et al. using (2.16) and the novel improved version based on (2.28). Furthermore, only the 3D case is considered subsequently since the 2D case is trivial (see Remark 2.6).

2.C.1. Approximation of the deformation gradient

In this Section we will show that requirement $\mathbf{F}^{\mathrm{h},e} = \mathbf{F}_0$ is fulfilled by all three versions of the deformation gradient considered in this work. We start with the isoparametric concept, where in view of (2.14), (2.15) and (2.13) we get for the compatible deformation gradient

$$\begin{aligned}\mathbf{F}^{\mathrm{h},e} &= \sum_{I=1}^{8} \boldsymbol{\varphi}_I^e \otimes \nabla_{\mathbf{X}} N_I = \mathbf{F}_0 \sum_{I=1}^{8} \left(\mathbf{X}_I^e \otimes \nabla_\xi N_I\right) (\mathbf{J}^{\mathrm{h},e})^{-1} + \mathbf{c} \otimes \nabla_\xi \sum_{I=1}^{8} (N_I)\, (\mathbf{J}^{\mathrm{h},e})^{-1} \\ &= \mathbf{F}_0 \mathbf{J}^{\mathrm{h},e} (\mathbf{J}^{\mathrm{h},e})^{-1} + \mathbf{c} \otimes \nabla_\xi (1) (\mathbf{J}^{\mathrm{h},e})^{-1} = \mathbf{F}_0 \end{aligned} \tag{2.85}$$

which is the desired result. If we use the modified gradient (2.16) instead of (2.15) the proof gets more complicated. However, with similar steps as above we get

$$\begin{aligned}\mathbf{F}^{\mathrm{h},e} &= \sum_{I=1}^{8} \boldsymbol{\varphi}_I^e \otimes \tilde{\nabla}_{\mathbf{X}} N_I = \mathbf{F}_0 \sum_{I=1}^{8} \left(\mathbf{X}_I^e \otimes \tilde{\nabla}_{\mathbf{X}} N_I\right) + \mathbf{c} \otimes \sum_{I=1}^{8} \tilde{\nabla}_{\mathbf{X}} N_I \\ &= \mathbf{F}_0 \sum_{I=1}^{8} \left[\mathbf{X}_I^e \otimes \left(\nabla_\xi N_I|_{\xi=0} + \sum_{A=1}^{4} \frac{j_0}{j^{\mathrm{h},e}} \nabla_\xi H_A \gamma_I^A \right)\right] \mathbf{J}_0^{-1} \\ &= \mathbf{F}_0 + \mathbf{F}_0 \sum_{I=1}^{8} \left[\mathbf{X}_I^e \otimes \sum_{A=1}^{4} \frac{j_0}{j^{\mathrm{h},e}} \nabla_\xi H_A \gamma_I^A \right] \mathbf{J}_0^{-1} \end{aligned} \tag{2.86}$$

where use has been made of identity

$$\begin{aligned}\sum_{I=1}^{8} \tilde{\nabla}_{\mathbf{X}} N_I &= \sum_{I=1}^{8} \nabla_0 N_I + \frac{j_0}{j^{\mathrm{h},e}} \mathbf{J}_0^{-\mathrm{T}} \sum_{I=1}^{8} \sum_{A=1}^{4} \nabla_\xi H_A \gamma_I^A \\ &= \frac{j_0}{8 j^{\mathrm{h},e}} \mathbf{J}_0^{-\mathrm{T}} \left[\sum_{A=1}^{4} \sum_{J=1}^{8} \nabla_\xi H_A \left(h_J^A \mathbf{X}_J^e \cdot \sum_{I=1}^{8} \nabla_0 N_I \right) + \sum_{I=1}^{8} \sum_{A=1}^{4} \nabla_\xi H_A h_I^A \right] = \mathbf{0}. \end{aligned} \tag{2.87}$$

Note, that the last equation can be established with help of $\sum_{I=1}^{8} \nabla_\xi N_I = \mathbf{0}$ and definition (2.20) of the gamma stabilization vectors. Since the desired result is $\mathbf{F}^{\mathrm{h},e} = \mathbf{F}_0$, the second term in the last line of (2.86) has to vanish. With definition (2.20) of the gamma stabilization vectors we get after basic, but lengthy tensor algebra

$$\sum_{I=1}^{8} \left[\mathbf{X}_I^e \otimes \sum_{A=1}^{4} \nabla_\xi H_A \gamma_I^A \right] = \frac{1}{8} \sum_{I=1}^{8} \sum_{A=1}^{4} \left(\mathbf{X}_I^e h_I^A \otimes \nabla_\xi H_A - \mathbf{J}_0 \mathbf{J}_0^{-1} \mathbf{X}_I^e h_I^A \otimes \nabla_\xi H_A \right) = \mathbf{0}. \tag{2.88}$$

Thus, $\mathbf{F}^{\mathrm{h},e} = \mathbf{F}_0$ is established for the modification of Simo et al.[132] as well. For the novel, improved version of that modification given in (2.28) we start by noting relations

$$\sum_{I=1}^{8} \mathbf{X}_I^e \otimes \bar{\nabla}_{\mathbf{X}} N_I = \mathbf{I} \qquad \text{and} \qquad \sum_{I=1}^{8} \bar{\nabla}_{\mathbf{X}} N_I = \mathbf{0} \tag{2.89}$$

which follow, albeit with tedious algebra, from the definition of the average gradient of shape functions (2.27). A proof of the last identities can, e.g., be found in Flanagan and Belytschko [40]. With this information at hand and similar steps as above we get

$$\mathbf{F}^{\mathrm{h},e} = \sum_{I=1}^{8} \boldsymbol{\varphi}_I^e \otimes \hat{\nabla}_{\mathbf{X}} N_I = \mathbf{F}_0 \sum_{I=1}^{8} \left(\mathbf{X}_I^e \otimes \hat{\nabla}_{\mathbf{X}} N_I \right) + \mathbf{c} \otimes \sum_{I=1}^{8} \hat{\nabla}_{\mathbf{X}} N_I = \mathbf{F}_0. \tag{2.90}$$

Thus, we have shown that criterion $\mathbf{F}^{\mathrm{h},e} \equiv \mathbf{F}_0$ holds for all versions of the compatible deformation gradient considered.

2.C.2. Finite element residual

The result of the previous section shows that the same constant state of strain is obtained regardless of the gradient of shape functions used. Together with the assumption of a homogeneous material this yields analogously the same constant stress $\mathbf{P}^{\mathrm{h},e} = \mathbf{P}_0$ for all three versions. With this information at hand, the finite element residual of one finite element with isoparametric formulation using the work-conjugate measures $\mathbf{P}$ and $\mathbf{F}$ is given by

$$\delta_{\varphi}\Pi = \int_{\Omega^e} \mathbf{P}^{\mathrm{h},e} : \delta_{\varphi}\mathbf{F}^{\mathrm{h},e} \, \mathrm{d}V = \sum_{I=1}^{8} \mathbf{P}_0 \delta\boldsymbol{\varphi}_I \cdot \int_{\Omega^e} \nabla_{\mathbf{X}} N_I \, \mathrm{d}V. \tag{2.91}$$

Note, that the equation above is exact, in the sense that it coincides with the analytic solution, because the isoparametric element fulfills the patch test. In the following we compare the residuals of the other two versions of the deformation gradient to that result.

Starting with the modification (2.16) of the gradient of shape functions by Simo et al. we get with $\int_{\hat{\Omega}} \nabla_{\xi} H_A \, \mathrm{d}\hat{\Omega} = \mathbf{0}, \; A = 1, \ldots, 4$ and straightforward steps

$$\begin{aligned}
\delta_{\varphi}\tilde{\Pi} &= \sum_{I=1}^{8} \mathbf{P}_0 \delta\boldsymbol{\varphi}_I \cdot \int_{\Omega^e} \tilde{\nabla}_{\mathbf{X}} N_I \, \mathrm{d}V \\
&= \sum_{I=1}^{8} \mathbf{P}_0 \delta\boldsymbol{\varphi}_I \cdot \left[\int_{\Omega^e} \nabla_0 N_I \, \mathrm{d}V + j_0 \mathbf{J}_0{}^{-\mathrm{T}} \sum_{A=1}^{4} \int_{\hat{\Omega}} \frac{1}{j^{\mathrm{h},e}} \nabla_{\xi} H_A j^{\mathrm{h},e} \, \mathrm{d}\hat{\Omega} \, \gamma_I^A \right] \\
&= \sum_{I=1}^{8} \mathbf{P}_0 \delta\boldsymbol{\varphi}_I \cdot \int_{\Omega^e} \nabla_0 N_I \, \mathrm{d}V.
\end{aligned} \tag{2.92}$$

Since (2.91) and (2.92) are in general not the same, the patch test is not fulfilled for generally distorted meshes. However, in case of a initially parallelepiped-shaped element $\int_{\Omega^e} \nabla_0 N_I \, \mathrm{d}V = \int_{\Omega^e} \nabla_X N_I \, \mathrm{d}V$ holds and the patch test is therefore fulfilled in that case. More favorably is the behavior of the improved version of the modification by Simo et al. given in (2.28). With definition (2.27) the residual reads

$$
\begin{aligned}
\delta_{\varphi}\tilde{\Pi} &= \sum_{I=1}^{8} \mathbf{P}_0 \delta\boldsymbol{\varphi}_I \cdot \int_{\Omega^e} \hat{\nabla}_X N_I \, \mathrm{d}V \\
&= \sum_{I=1}^{8} \mathbf{P}_0 \delta\boldsymbol{\varphi}_I \cdot \left[\frac{1}{V} \int_{\Omega^e} \nabla_X N_I \, \mathrm{d}V \int_{\Omega^e} \mathrm{d}V + j_0 \mathbf{J}_0^{-\mathrm{T}} \sum_{A=1}^{4} \int_{\hat{\Omega}} \frac{1}{j^{\mathrm{h},e}} \nabla_{\xi} H_A j^{\mathrm{h},e} \, \mathrm{d}\hat{\Omega} \, \hat{\gamma}_I^A \right] \\
&= \sum_{I=1}^{8} \mathbf{P}_0 \delta\boldsymbol{\varphi}_I \cdot \int_{\Omega^e} \nabla_X N_I \, \mathrm{d}V.
\end{aligned}
\tag{2.93}
$$

which is exactly the same as (2.91). Thus, the patch test is fulfilled for arbitrary initial element shapes if the novel approach presented in this work is used.

3. Improving efficiency and robustness of EAS elements for nonlinear problems

This chapter reproduces:*
Pfefferkorn R, Bieber S, Oesterle B, Bischoff M, and Betsch P. "Improving Efficiency and Robustness of EAS Elements for Nonlinear Problems". In: *Int J Numer Meth Eng.* 122(8): 1911–1939, 2021. DOI: 10.1002/nme.6605

Abstract: The enhanced assumed strain (EAS) method is one of the most frequently used methods to avoid locking in solid and structural finite elements. One issue of EAS elements in the context of geometrically non-linear analyses is their lack of robustness in the Newton-Raphson scheme, which is characterized by the necessity of small load increments and large numbers of iterations. In the present work we extend the recently proposed mixed integration point (MIP) method to EAS elements in order to overcome this drawback in numerous applications. Furthermore, the MIP method is generalized to generic material models, which makes this simple method easily applicable for a broad class of problems. In the numerical simulations in this work we compare standard strain based EAS elements and their MIP improved versions to elements based on the assumed stress method in order to explain when and why the MIP method allows to improve robustness. A further novelty in the present work is an inverse stress-strain relation for a Neo-Hookean material model.

Keywords: enhanced assumed strain (EAS), inverse stress–strain relation, mixed finite elements, mixed integration point method, Newton–Raphson scheme, robustness

3.1. Introduction

In the early days of the finite element method (FEM), it was soon discovered that low-order purely displacement-based (U) finite elements yield poor results in many cases due to severe locking phenomena. Therefore, a plethora of mixed finite elements has been developed. Two of the probably most successful classes in linear analyses are *enhanced assumed strain* (EAS) elements introduced in 1990 by Simo and Rifai [134] as a generalization of the popular *incompatible modes* elements by Taylor et al. [141] and *assumed stress* (AS) elements proposed in the 1980s by Pian and Sumihara [110] and Pian and Tong [112] for 2D and 3D problems, respectively (see also the pioneering work of Fraeijs de Veubeke [42]). Both classes exhibit excellent behavior in linear simulations. They are completely locking-free if the additional fields are approximated appropriately, they are stable and they are relatively insensitive to mesh distortion. In fact, there are hardly any drawbacks of using such elements in the linear elastic case. Furthermore, it is shown by Bischoff et al. [21] that for every EAS element an equivalent AS element can be found and vice versa.

For nonlinear problems, however, the two approaches are not identical anymore. Here, AS elements are less popular, since they need an *inverse stress-strain relation*. With a few exceptions for simple material models (see, e.g., Wriggers [154]), this is extremely difficult or even impossible to obtain. In cases, where a solution exists, the recent numerical approach by Viebahn et al. [145] is a simple alternative to the cumbersome analytical inversion. However, there exist cases in which inversion of the stress strain relation is impossible (see, e.g., Ogden [98] and the present work). EAS elements do not encounter these difficulties, because their strain based construction naturally fits the typically strain driven format of non-linear material laws. Thus, they can easily be extended to general material models, also including geometric nonlinearity, see for example [8, 30, 31, 46, 47, 66, 67, 69, 104, 105, 131, 132, 155], among others. Unfortunately, two major open issues of EAS elements remain. First, unphysical instabilities occur, depending on the chosen material model, ansatz functions for the enhanced field and state of strain. This was first discovered by Reese and Wriggers [156] and addressed in many subsequent works (notably [8, 47, 69]). The effect is also present for AS element and other locking-free formulations. The second open issue is lack of *robustness* in the *Newton-Raphson* (NR) solution algorithm. By the term robustness we herein refer to two properties: maximum size of applicable load steps and number of NR iterations required to find equilibrium. In this sense, robust elements are also efficient, since they require fewer iterations and therefore also fewer time consuming matrix factorizations. In this paper, only the second open issue is addressed, which seems to be a particular feature of EAS elements (AS elements, for instance, are superior in this respect).
So far, this kind of algorithmic robustness has received little attention in the development of finite elements and focus has been put onto other topics, such as locking and stability. In the context of EAS elements the only works known to the authors that deal with robustness are [65, 69, 145, 146, 151, 152]. For beam finite elements Garcea et al. [45] discovered that (AS-type) mixed finite elements provide superior robustness compared to displacement-based elements. This result has recently been reproduced by Magisano et al. [87] for solid-shell problems. In the context of continuum elements Viebahn et al. [145] obtained similar results while comparing AS and EAS elements. The authors showed that EAS elements have inferior robustness even if they yield the same equilibrium solution as AS elements. For another class of mixed finite elements, which are based on a Hu-Washizu type variational functional, superior robustness of the mixed format compared to the displacement form has been observed by Wisniewski et al. [151, 152] and Betsch et al. [19].

One approach to improve the behavior of strain-driven elements is presented by Mei et al. [91] who propose to transform nonlinear finite element equations to make them more "linear". A simple example for this would be transforming the equation $e^x = a$ with the unknown x and a constant a by taking the logarithm on both sides. Unfortunately, this mathematically elegant approach is difficult to implement into practical simulations, since a suitable transformation is not straightforward to determine. Moreover, the main source of nonlinearity has to be known a priori, which becomes difficult if multiple sources of nonlinearity (e.g., geometric and material nonlinearities, constraints, plasticity) are present or if their corresponding impact changes during simulation.
In the context of shell problems, Magisano et al. [88] introduce the *mixed integration point* (MIP) method (see also [43, 89]). The key idea is to introduce independent stresses at the Gauss quadrature points, which ultimately leads to a modification of the stresses used for the geometric stiffness matrix. A special feature of the method is that the residuum is not altered, which means that only the robustness of the method during iteration is (usually) improved without changing the converged result. The principal idea of modifying the stress in the geometric part of the stiffness matrix has already been proposed in the 1980s by Kuo-Mo et al. [72, 73]. They suggested to use an inconsistent tangent, by using the stress of the previous converged load step for all Newton-Raphson iterations of the next load step in order to compute the geometric stiffness contributions. However, the MIP update algorithm is more sophisticated and improves upon those results.
In the context of contact problems, Zavarise et al. [160] proposed a similar philosophy of modifying the geometric stiffness contributions as a way of efficiently dealing with large penetrations and increasing the robustness of contact algorithms.

In the present work we use the idea of the MIP method introduced by Magisano et al. [88] and apply it to solid finite elements instead of structural (solid-)shell and beam elements, which are usually considered in the literature [88, 89]. Furthermore, we propose three extensions to the MIP method. First of all, we consider general material models, since being restricted to the simple St. Venant-Kirchhoff material model is one of the current limitations of the MIP method. We discuss under which conditions the MIP method can be applied successfully to more advanced materials and when less beneficial results are to be expected. Secondly, we propose a transformation of the method to use spatial stress and strain measures in order to support computationally more efficient elements and allow simple implementation of many material models. Finally, and most importantly, we apply the method to EAS elements, which significantly increases their robustness and makes this class of elements even more interesting in practical simulations.
Various numerical examples, including elastic and plastic material laws, are presented to highlight the properties and performance of the MIP method. With these we show, when and why the method allows the greatest benefit and when only little improvement is to be expected.
A final novelty of this work is an analytic inverse stress-strain relation for a Neo-Hookean material model, which can be used for AS elements. It demonstrates the limits of this class of elements in general nonlinear simulations.

The present work is structured into seven sections. General relations and notation for nonlinear continuum mechanics and finite element formulations of the displacement-based (U) as well as EAS and AS mixed finite elements are given in Section 3.2. Afterwards, a simple model problem is presented in Section 3.3 to show basic relations exploited for the MIP method. The MIP method and its extension to general material models are described in Section 3.4. Additional extensions and more details of the MIP method are given in Section 3.5. In particular, Section 3.5.1 covers the MIP method for EAS elements while sections 3.5.2 to 3.5.4 deal with its application to problems based on spatial quantities, the implementation of the method and a simple one-element test. Extensive numerical simulations follow in Section 3.6 before conclusions are drawn in Section 3.7. Appendix 3.A covers the St. Venant-Kirchhoff, Neo-Hookean and elasto-plastic material models used throughout this work. Special emphasis is put on the invertibility of the stress-strain relation of the Neo-Hookean model (see Section 3.A.2.2). More details concerning the simple model problem in Section 3.3 are given in Appendix 3.B.

3.2. Mixed finite elements for solid mechanics

3.2.1. Continuum mechanics

The deformation of a deformable body from its reference configuration $\mathcal{B}_0$ to the current configuration $\mathcal{B}$ is described by the bijective deformation map

$$\boldsymbol{\varphi} \in \mathcal{U} = \left\{ \boldsymbol{\varphi} : \mathcal{B}_0 \to \mathbb{R}^3 \,\middle|\, (\boldsymbol{\varphi})_i \in H_1,\ \det(\mathrm{D}\boldsymbol{\varphi}) > 0 \text{ and } \boldsymbol{\varphi}(\mathbf{X}) = \bar{\boldsymbol{\varphi}}(\mathbf{X}),\ \mathbf{X} \in \partial_\varphi \mathcal{B}_0 \right\}, \quad (3.1)$$

which maps material points $\mathbf{X} \in \mathcal{B}_0$ to corresponding spatial points $\mathbf{x} = \boldsymbol{\varphi}(\mathbf{X}) \in \mathcal{B}$. Therein, $\partial_\varphi \mathcal{B}_0$ denotes the part of the body's boundary $\partial \mathcal{B}_0$ on which the deformations are prescribed by $\bar{\boldsymbol{\varphi}} : \partial_\varphi \mathcal{B}_0 \to \mathbb{R}^3$. Linearization of the deformation map[1] $\boldsymbol{\varphi}$ at a point $\mathbf{X}$ yields the *deformation gradient*

$$\mathbf{F}_\varphi(\mathbf{X}) = \frac{\partial \boldsymbol{\varphi}}{\partial \mathbf{X}} = \mathrm{D}\boldsymbol{\varphi}, \quad (3.2)$$

where the index φ denotes that $\mathbf{F}_\varphi$ is computed from the deformations alone, which will not necessarily be the case in subsequent sections. The deformation gradient[2] $\mathbf{F}$ can be used to define the *right Cauchy-Green tensor* and *Green-Lagrange strain tensor*, respectively, as

$$\mathbf{C} = \mathbf{F}^{\mathrm{T}}\mathbf{F}, \qquad \mathbf{E} = \frac{1}{2}\left(\mathbf{F}^{\mathrm{T}}\mathbf{F} - \mathbf{I}\right). \quad (3.3)$$

[1] Subsequently, arguments of functions are frequently omitted in order to improve readability and notational simplicity.

[2] Index φ is omitted here since the following relations are valid for general deformation gradients.

Furthermore, the constitutive *second Piola-Kirchhoff stress tensor* and its linearization with respect to E are introduced as

$$\hat{\mathrm{S}} = \hat{\mathrm{S}}(\mathbf{E}, \Xi), \qquad \Delta_E \hat{\mathrm{S}} = \hat{\mathbb{C}} : \Delta \mathbf{E}. \tag{3.4}$$

The notation $\Delta_{(\bullet)}(*)$ is used to denote linearization of $(*)$ with respect to $(\bullet)$. The form (3.4) of constitutive law for $\hat{\mathrm{S}}$ covers a wide range of material models including inelastic behavior via internal variables Ξ. Linearization of the constitutive second Piola-Kirchhoff stress tensor with respect to E is governed by the fourth-order material tensor $\hat{\mathbb{C}} = \partial \hat{\mathrm{S}} / \partial \mathbf{E}$.

Remark 3.1. *In case of a homogeneous* hyperelastic *material with strain energy function* $W(\mathbf{E})$, *relations* (3.4) *are given by*

$$\hat{\mathrm{S}} = \frac{\partial W}{\partial \mathbf{E}}, \qquad \hat{\mathbb{C}} = \frac{\partial^2 W}{\partial \mathbf{E} \partial \mathbf{E}}. \tag{3.5}$$

However, we emphasize that hyperelasticity is no prerequisite for the novel techniques proposed in this work.

3.2.2. Finite element method

3.2.2.1. Displacement-based finite elements

In the sequel, we only consider numerical solutions of nonlinear solid mechanic problems obtained with the *finite element method* (FEM). More specifically, a body $\mathcal{B}$ is approximated with n_{el} four-node quadrilateral or eight-node brick finite elements Ω^e in 2D and 3D respectively. Furthermore, the isoparametric concept is applied. Approximations of geometry $\mathbf{X}$ and deformation $\boldsymbol{\varphi}$ within one element Ω^e are then given by

$$\mathbf{X}^{\mathrm{h},e} = \sum_{I=1}^{n_{\mathrm{no}}} N_I(\xi) \mathbf{X}_I^e, \qquad \boldsymbol{\varphi}^{\mathrm{h},e} = \sum_{I=1}^{n_{\mathrm{no}}} N_I(\xi) \boldsymbol{\varphi}_I^e, \tag{3.6}$$

where superscript h denotes an approximation and e marks elementwise quantities. However, these superscripts are usually omitted in the remainder of this work since most of the subsequent presentations are dealing with discretized fields on element level. Thus, if not specified otherwise, every quantity introduced in the sequel is an approximation on element level, for example $\boldsymbol{\varphi} \equiv \boldsymbol{\varphi}^{\mathrm{h},e}$. In the equation above, n_{no} is the number of nodes of the element and $\mathbf{X}_I^e$, $\boldsymbol{\varphi}_I^e$ are the nodal values of the respective quantities. Moreover, $N_I(\xi)$ denote the *Lagrangian* shape functions defined on reference element $\hat{\Omega} = [-1, 1]^{n_{\mathrm{dim}}}$ for two- and three-dimensional approximations, respectively (see, e.g., [154]).

The *internal* part of the weak form of equilibrium of a purely displacement-based formulation on element level is given by (see, e.g., [25, 154])

$$G_{\mathrm{d},\varphi}(\boldsymbol{\varphi}, \delta\boldsymbol{\varphi}) = \int_{\Omega^e} \hat{\mathrm{S}}_\varphi : \delta_{\boldsymbol{\varphi}} \mathbf{E}_\varphi \, \mathrm{d}V, \tag{3.7}$$

where $\delta\boldsymbol{\varphi}$ is a kinematically admissible test function, approximated in the same way as $\boldsymbol{\varphi}$ in (3.6) (*Bubnov-Galerkin* approach) and $\delta_{(\bullet)}(*)$ denotes the first variation of $(*)$ with respect to $(\bullet)$. The displacement-based Green-Lagrange tensor $\mathbf{E}_\varphi$ and constitutive second Piola-Kirchhoff stress $\hat{\mathbf{S}}_\varphi$ are defined through (3.2), (3.3) and (3.4) as $\mathbf{E}_\varphi = \mathbf{E}(\mathbf{F}_\varphi)$ and $\hat{\mathbf{S}}_\varphi = \hat{\mathbf{S}}(\mathbf{E}_\varphi, \Xi)$, respectively. In these relations, index φ denotes the purely displacement-based nature of these quantities and index $\mathbb{d}$ specifies that the respective variable is part of the displacement-based FEM.[3] The *external* part $G_\varphi^{\text{ext}}(\boldsymbol{\varphi}, \delta\boldsymbol{\varphi})$ of the weak form includes all external forces acting on $\mathcal{B}_0$ and the boundary $\partial_t\mathcal{B}_0 = \partial\mathcal{B}_0 \setminus \partial_\varphi\mathcal{B}_0$. They are not specified in detail here, since only the internal part (3.7) will be needed for the developments in this paper.

Considering the arbitrariness of $\delta\boldsymbol{\varphi}$ and using Gauss quadrature to approximate the integral in (3.7), the linearized discrete form of (3.7) on element level reads

$$\sum_{g=1}^{n} \left[\mathbf{M}_{\mathbb{d},g}^{\varphi\varphi} + \mathbf{G}_{\mathbb{d},g}^{\varphi\varphi}(\hat{\mathbf{S}}_{\varphi,g})\right] w_g \Delta\boldsymbol{\varphi}^e = -\sum_{g=1}^{n} \mathbf{R}_{\mathbb{d},g}^{\varphi} w_g, \tag{3.8}$$

where

$$\mathbf{R}_{\mathbb{d},g}^{\varphi} = \left[(\mathbf{B}_{\mathbb{d}}^{\varphi})^{\mathrm{T}}\hat{\mathbf{S}}_\varphi\right]_g, \tag{3.9a}$$

$$\mathbf{M}_{\mathbb{d},g}^{\varphi\varphi} = \left[(\mathbf{B}_{\mathbb{d}}^{\varphi})^{\mathrm{T}}\hat{\mathbb{C}}_\varphi\mathbf{B}_{\mathbb{d}}^{\varphi}\right]_g, \tag{3.9b}$$

$$\mathbf{G}_{\mathbb{d},g}^{\varphi\varphi} = \mathbf{G}_{\mathbb{d}}^{\varphi\varphi}(\hat{\mathbf{S}}_{\varphi,g}), \tag{3.9c}$$

denote the integrand of the element residual as well as the integrand of the material and geometric part of the stiffness matrix, evaluated at the integration points (for more details see, e.g., Wriggers[154] Ch. 4.2.2). Abbreviation $(\bullet)_g = (\bullet)(\boldsymbol{\xi}_g)$ denotes evaluation of a quantity at Gauss point $g = 1, \ldots, n$ with weight w_g. Quantities $\hat{\mathbf{S}}_\varphi$ and $\hat{\mathbb{C}}_\varphi$ in (3.9a) and (3.9b) have to be cast in *Voigt* (vector-matrix) notation, which is not separately marked since it becomes clear from the context when this notation is required. Superscripts in (3.8) and (3.9), indicating applied directional derivatives w.r.t. φ, are not necessary here but are merely introduced to get similar notation as for the mixed formulations presented in the sequel. The nodal operator matrix of the material displacement form is denoted by $\mathbf{B}_{\mathbb{d}}^{\varphi}$. We emphasize the dependence of $\mathbf{G}_{\mathbb{d},g}^{\varphi\varphi}$ on the constitutive stress $\hat{\mathbf{S}}_{\varphi,g}$ in (3.8), since this is the key to the MIP method presented in Sections 3.4 and 3.5.

[3] The difference between index φ and $\mathbb{d}$ becomes apparent in Section 3.2.2.2, where the deformation-based deformation gradient $\mathbf{F}_\varphi$ is needed as well. However, the weak form and its approximation can not be governed with the same quantities as presented in this section, which is why index $\mathbb{d}$ is introduced.

3.2.2.2. EAS method

The first well-known class of multi-field finite elements used in this work are EAS elements which were first proposed for nonlinear problems by Simo and Armero [131]. In the present work, we give only a brief overview and refer to Pfefferkorn and Betsch [104], among others, for a more thorough summary of many topics concerning EAS elements.

The key idea of the EAS method in the form used herein is to recast the deformation gradient in the form

$$\mathbf{F}_{\alpha}(\boldsymbol{\varphi}, \boldsymbol{\alpha}) = \mathbf{F}_{\varphi}(\boldsymbol{\varphi}) + \tilde{\mathbf{F}}(\boldsymbol{\varphi}, \boldsymbol{\alpha}), \tag{3.10}$$

where $\mathbf{F}_{\varphi}$ and $\tilde{\mathbf{F}}$ denote the compatible and incompatible (or enhanced) part of the deformation gradient. The latter includes additional enhanced degrees of freedom $\boldsymbol{\alpha}$, which are used to improve the element's behavior with respect to locking.

The variational basis for EAS finite elements is a *Hu-Washizu* [148] type functional. Usually, the independent stress is eliminated by enforcing the discrete stress and the enhanced strain field to be L_2-orthogonal. This ultimately yields the approximated weak form[4] formulated using material quantities S and E in the form

$$G_{\mathbb{e},\varphi} = \int_{\Omega^e} \hat{\mathbf{S}}_{\alpha} : \delta_{\boldsymbol{\varphi}} \mathbf{E}_{\alpha} \,\mathrm{d}V, \tag{3.11a}$$

$$G_{\mathbb{e},\alpha} = \int_{\Omega^e} \hat{\mathbf{S}}_{\alpha} : \delta_{\boldsymbol{\alpha}} \mathbf{E}_{\alpha} \,\mathrm{d}V = 0, \tag{3.11b}$$

where the Green-Lagrange strain (3.3) and constitutive second Piola-Kirchhoff stress (3.4) are given by $\mathbf{E}_{\alpha} = \mathbf{E}(\mathbf{F}_{\alpha})$ and $\hat{\mathbf{S}}_{\alpha} = \hat{\mathbf{S}}(\mathbf{E}_{\alpha}, \Xi)$, respectively. On top of that, index $\mathbb{e}$ is introduced to mark quantities of the EAS method in analogy to index $\mathbb{d}$ used in Section 3.2.2.1 for the displacement-based method. Note that in (3.11) only (3.11b) is equal to zero on element level without global assembly since the enhanced degrees of freedom $\boldsymbol{\alpha}$ are introduced elementwise.

Approximation of the deformation $\boldsymbol{\varphi}$ and compatible deformation gradient is applied as described in Section 3.2.2.1 and the enhanced deformation gradient can be discretized on element level using *Wilson*-modes [149] with nine (in 3D) enhanced parameters α_i by

$$\tilde{\mathbf{F}} = \mathbf{F}_0 \frac{j_0}{j} \mathbf{J}_0^{-\mathrm{T}} \begin{bmatrix} \xi\alpha_1 & \eta\alpha_2 & \zeta\alpha_3 \\ \xi\alpha_4 & \eta\alpha_5 & \zeta\alpha_6 \\ \xi\alpha_7 & \eta\alpha_8 & \zeta\alpha_9 \end{bmatrix} \mathbf{J}_0^{-1}. \tag{3.12}$$

In the last equation, $\mathbf{J}_0$ and j_0 are the evaluation of the Jacobian of the isoparametric map $\mathbf{J}$ and its determinant $j = \det(\mathbf{J})$ at the element centroid $\boldsymbol{\xi} = \mathbf{0}$. Moreover, $\mathbf{F}_0 = \mathbf{F}_{\varphi}(\boldsymbol{\xi} = \mathbf{0})$ denotes the compatible deformation gradient at the element center and is needed to ensure objectivity of the formulation.

---

[4] This is essentially equivalent to the stationary conditions of the Hu-Washizu functional, but it is not limited to hyperelastic behavior.

Remark 3.2. *We refer to Pfefferkorn and Betsch [104], among others, for many other possibilities to approximate both, the compatible and the incompatible part of the deformation gradient and for a thorough explanation of the requirements to be met. Furthermore, an overview of other ansatz functions for the enhanced deformation gradient is given in the aforementioned reference. Note especially the transposed Wilson-modes which cure the spurious hourglass instability of the standard EAS element in compression.*

The discrete linearized form of (3.11) in matrix notation is given by

$$\sum_{g=1}^{n} w_g \begin{bmatrix} \mathbf{M}_{\mathbb{e},g}^{\varphi\varphi} + \mathbf{G}_{\mathbb{e},g}^{\varphi\varphi}(\hat{\mathbf{S}}_{\alpha,g}) & \mathbf{M}_{\mathbb{e},g}^{\varphi\alpha} + \mathbf{G}_{\mathbb{e},g}^{\varphi\alpha}(\hat{\mathbf{S}}_{\alpha,g}) \\ \mathbf{M}_{\mathbb{e},g}^{\alpha\varphi} + \mathbf{G}_{\mathbb{e},g}^{\alpha\varphi}(\hat{\mathbf{S}}_{\alpha,g}) & \mathbf{M}_{\mathbb{e},g}^{\alpha\alpha} + \mathbf{G}_{\mathbb{e},g}^{\alpha\alpha}(\hat{\mathbf{S}}_{\alpha,g}) \end{bmatrix} \begin{bmatrix} \Delta\boldsymbol{\varphi}^e \\ \Delta\boldsymbol{\alpha}^e \end{bmatrix} = -\sum_{g=1}^{n} w_g \begin{bmatrix} \mathbf{R}_{\mathbb{e},g}^{\varphi} \\ \mathbf{R}_{\mathbb{e},g}^{\alpha} \end{bmatrix}, \tag{3.13}$$

where the integrands of the element residuals and the geometric as well as material parts of the tangent are given by

$$\mathbf{R}_{\mathbb{e},g}^{\varphi} = \left[(\mathbf{B}_{\mathbb{e}}^{\varphi})^{\mathrm{T}}\hat{\mathbf{S}}_{\alpha}\right]_g, \qquad \mathbf{R}_{\mathbb{e},g}^{\alpha} = \left[(\mathbf{B}_{\mathbb{e}}^{\alpha})^{\mathrm{T}}\hat{\mathbf{S}}_{\alpha}\right]_g, \tag{3.14a}$$

$$\mathbf{M}_{\mathbb{e},g}^{\varphi\varphi} = \left[(\mathbf{B}_{\mathbb{e}}^{\varphi})^{\mathrm{T}}\hat{\mathbb{C}}_{\alpha}\mathbf{B}_{\mathbb{e}}^{\varphi}\right]_g, \qquad \mathbf{G}_{\mathbb{e},g}^{\varphi\varphi} = \mathbf{G}_{\mathbb{e}}^{\varphi\varphi}(\hat{\mathbf{S}}_{\alpha,g}), \tag{3.14b}$$

$$\mathbf{M}_{\mathbb{e},g}^{\varphi\alpha} = \left[(\mathbf{B}_{\mathbb{e}}^{\varphi})^{\mathrm{T}}\hat{\mathbb{C}}_{\alpha}\mathbf{B}_{\mathbb{e}}^{\alpha}\right]_g, \qquad \mathbf{G}_{\mathbb{e},g}^{\varphi\alpha} = \mathbf{G}_{\mathbb{e}}^{\varphi\alpha}(\hat{\mathbf{S}}_{\alpha,g}), \tag{3.14c}$$

$$\mathbf{M}_{\mathbb{e},g}^{\alpha\varphi} = \left[(\mathbf{B}_{\mathbb{e}}^{\alpha})^{\mathrm{T}}\hat{\mathbb{C}}_{\alpha}\mathbf{B}_{\mathbb{e}}^{\varphi}\right]_g, \qquad \mathbf{G}_{\mathbb{e},g}^{\alpha\varphi} = \mathbf{G}_{\mathbb{e}}^{\alpha\varphi}(\hat{\mathbf{S}}_{\alpha,g}), \tag{3.14d}$$

$$\mathbf{M}_{\mathbb{e},g}^{\alpha\alpha} = \left[(\mathbf{B}_{\mathbb{e}}^{\alpha})^{\mathrm{T}}\hat{\mathbb{C}}_{\alpha}\mathbf{B}_{\mathbb{e}}^{\alpha}\right]_g, \qquad \mathbf{G}_{\mathbb{e},g}^{\alpha\alpha} = \mathbf{G}_{\mathbb{e}}^{\alpha\alpha}(\hat{\mathbf{S}}_{\alpha,g}). \tag{3.14e}$$

Here, superscripts φ and α denote the various parts of strain-displacement matrices, tangent and residual. Moreover, $\hat{\mathbb{C}}_{\alpha}$ is the material tangent (3.4) evaluated using $\mathbf{E}_{\alpha}$. Like in the previous section, the geometric part of the stiffness matrix depends on the constitutive stress $\hat{\mathbf{S}}_{\alpha}$.

Remark 3.3. *Usually, static condensation is performed on element level to eliminate the internal degrees of freedom $\Delta\boldsymbol{\alpha}^e$ from* (3.13). *However, in this work the full form of* (3.13) *is needed for development of the MIP method, presented in Section 3.5.1, and static condensation is carried out afterwards.*

3.2.2.3. Assumed stress method

The second class of mixed finite elements considered in this work are *assumed stress* elements, which are based on a *Hellinger-Reissner* (HR) functional and employ separate stress approximation. However, with a few exceptions for simple material models, a complementary stored energy function, which is required for a HR-functional, does not exist. This is why usually a weak form is used as starting point for AS elements if more complex material models are to be used (see also [145, 154]). In the present work the discretized weak form is given by

$$G_{\mathfrak{s},\varphi} = \int_{\Omega^e} \mathbf{S}_\beta : \delta_\varphi \mathbf{E}_\varphi \, \mathrm{d}V, \tag{3.15a}$$

$$G_{\mathfrak{s},\beta} = \int_{\Omega^e} \delta_\beta \mathbf{S}_\beta : (\mathbf{E}_\varphi - \hat{\mathbf{E}}_\beta) \, \mathrm{d}V = 0, \tag{3.15b}$$

where the independent (not constitutive) stress tensor $\mathbf{S}_\beta$ is introduced as function of internal elementwise degrees of freedom $\boldsymbol{\beta}$ and index $\mathfrak{s}$ denotes the AS method. Common approximations for $\mathbf{S}_\beta$ can, among others, be found in the work of Viebahn et al. [145]. In this work, however, we apply a special stress approximation that is described in Section 3.4. The expression in (3.15b), like (3.11b), is zero on element level due to the elementwise approximations of $\mathbf{S}_\beta$. The constitutive strain $\hat{\mathbf{E}}_\beta$ introduced in (3.15) is computed from the inverse of relation (3.4) such that

$$\hat{\mathbf{E}}_\beta = \hat{\mathbf{E}}(\mathbf{S}_\beta, \Xi). \tag{3.16}$$

This requires that (3.4) is at least locally invertible, which is given for most commonly used material models in the neighborhood of the stress free reference configuration $\mathbf{F} = \mathbf{I}$, as described in Ogden [98] Ch. 6.2.2. The linearization of this relation can then be computed via the law of differentiation of the inverse by

$$\Delta_S \hat{\mathbf{E}}_\beta = \hat{\mathbb{D}}_\beta : \Delta \mathbf{S}_\beta, \tag{3.17}$$

where the tangential compliance matrix $\hat{\mathbb{D}}_\beta := \hat{\mathbb{C}}^{-1}(\hat{\mathbf{E}}_\beta, \Xi)$ is the inverse[5] of the material tangent (3.4).

Remark 3.4. *While most common material models are invertible in the neighborhood around the stress free configuration, analytic relations of inverse stress-strain relations rarely exist. Appendix 3.A covers analytic inversions. Besides the simply invertible St. Venant-Kirchhoff model, we present a novel inverse stress-strain relation for a Neo-Hookean model (see Appendix 3.A.2). More complex models are difficult or even impossible to analytically invert. Recently, a numerical procedure for general hyperelastic models has been proposed by Viebahn et al. [145].*

Remark 3.5. *Even though inversion is usually possible close to the stress free reference configuration, it is not necessarily so for states of larger strain. This is, e.g., the case for the Neo-Hookean model described in Appendix 3.A.2. The states of stress, for which unique invertibility is given for that model, are described in detail in Appendix 3.A.2.2. These requirements are a severe limitation of the applicability of AS elements since these states can be observed in practical simulations as shown in Section 3.6.3, where the non-invertibility leads to failure of the Newton-Raphson scheme.*

[5] The inverse of a fourth order tensor can easily be computed in Voigt notation where it reduces to computing the inverse of the corresponding matrix.

3.3. Motivation for the MIP method

3.3.1. Problem description

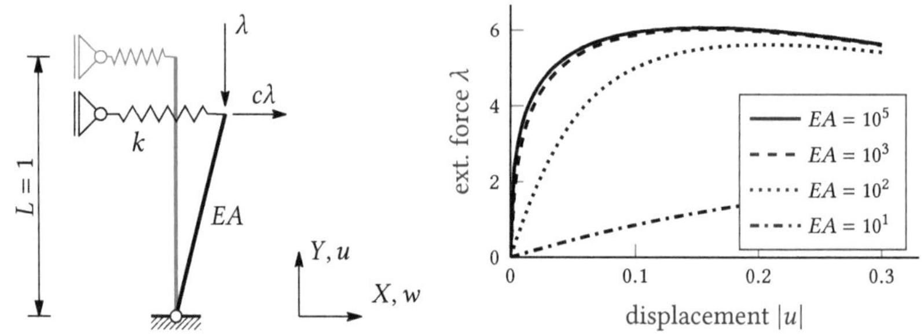

Figure 3.1.: Model problem. Setup and analytical load-displacement curves for $c = 0.25$, $k = 10$ and different values of the axial stiffness EA.

In this section, a simple model problem is considered to demonstrate issues that may arise in the iterative solution process and to motivate the MIP method presented in subsequent sections. Following Magisano et al.[87], the simple nonlinear bar-spring system, shown in Figure 3.1, serves as model problem. The structure consists of a geometrically nonlinear bar with axial stiffness EA and a linear spring with stiffness k. We use a unit length $L = 1$ for the bar in order to simplify expressions. The upper end of the bar is subjected to a vertical load λ and a horizontal load $c\lambda$. We summarize here the most important relations for this example. More details are given in Appendix 3.B. For a displacement-based formulation with two degrees of freedom $\mathbf{d}^{\mathrm{T}} = \begin{bmatrix} u & w \end{bmatrix}$, in analogy to Section 3.2.1, the kinematic and constitutive relations are

$$\varepsilon = u + \frac{1}{2}\left(u^2 + w^2\right), \qquad \hat{N} = EA\varepsilon, \tag{3.18}$$

where $\varepsilon(u, w)$ is the displacement-based axial Green-Lagrange strain and $\hat{N}(u, w)$ the constitutive "second Piola-Kirchhoff normal force". With these expressions at hand, the linearized displacement-based weak form of equilibrium (3.8) is given by

$$\begin{bmatrix} EA(1+u)^2 + \hat{N} & EA(1+u)w \\ EA(1+u)w & EAw^2 + k + \hat{N} \end{bmatrix} \begin{bmatrix} \Delta u \\ \Delta w \end{bmatrix} = -\begin{bmatrix} \hat{N}(1+u) + \lambda \\ \hat{N}w + kw - \lambda c \end{bmatrix}. \tag{3.19}$$

For the AS formulation, similar to Section 3.2.2.3, the linearized version of the weak form of kinematics *and* equilibrium (3.15) reads

$$\begin{bmatrix} -1/EA & 1+u & w \\ 1+u & N & 0 \\ w & 0 & k+N \end{bmatrix} \begin{bmatrix} \Delta N \\ \Delta u \\ \Delta w \end{bmatrix} = -\begin{bmatrix} \varepsilon - \hat{\varepsilon} \\ N(1+u) + \lambda \\ Nw + kw - \lambda c \end{bmatrix}, \tag{3.20}$$

where, for sake of simplicity, the independent "stress" field solely consists of the axial force N of the bar, whereas the linear spring remains unaffected. Furthermore, the constitutive strain $\hat{\varepsilon}$ is given by inverse stress-strain relation $\hat{\varepsilon} = N/EA$. Condensation of N reproduces equation (3.19). However, the performance of the incremental solution process can differ significantly if the matrices in (3.20) are used, which will be shown next.

For both formulations, the results after the first iteration ($j = 1$) of the standard Newton-Raphson procedure are considered, starting from a stress-free reference configuration $u^{(0)} = w^{(0)} = 0$, $N^{(0)} = 0$ and $\lambda^{(0)} = 0$. For both formulations, the first displacement predictor due to an incremental load $\Delta\lambda$ is the same,

$$\Delta u^{(0)} = -\frac{\Delta\lambda}{EA}, \qquad \Delta w^{(0)} = \frac{c\Delta\lambda}{k}. \tag{3.21}$$

While these values are identical for both formulations (which is not the case in general), the axial force may differ significantly. After the first iteration, the updated values for the constitutive and independent normal force are

$$\hat{N}^{(1)} = \hat{N}(u^{(1)}, w^{(1)}) = -\Delta\lambda + \frac{\Delta\lambda^2}{2EA} + \frac{\Delta\lambda^2 c^2 EA}{2k^2}, \qquad N^{(1)} = -\Delta\lambda, \tag{3.22}$$

which reveals the intrinsic problem of the extrapolated constitutive force $\hat{N}(u, w)$ compared to the independent field N. At equilibrium points both force values are identical, but during the iteration process this is not necessarily the case. The extra terms $\frac{\Delta\lambda^2}{2EA} + \frac{\Delta\lambda^2 c^2 EA}{2k^2}$ emanate from the fact that $\hat{N}$ is computed from the strain via the constitutive law. The strain, however, contains spurious contributions from the poor displacement predictor. For a high stiffness ratio EA/k^2, the extrapolated force $\hat{N}$ may be estimated far more inaccurately compared to the independent stress field N. In fact, $\hat{N}$ follows a quadratic path while N is extrapolated linearly. Since the extrapolated normal force affects the geometric part of the tangential stiffness matrix in the next iteration step ($j + 1$), the update of the NR scheme in this step will also be a worse guess in the displacement-based method. This ultimately explains why displacement-based finite elements may require more iterations within a NR scheme in comparison with AS elements, as demonstrated in Section 3.3.3.

As pointed out in [87], a similar phenomenon can occur in the structural analysis of thin walled structures, because of the high ratio of membrane stiffness to bending stiffness. This, especially for deformations involving large rotations (which is the case for the model problem herein), may spoil the extrapolated stress quantities tremendously. However, it is not relevant anymore when no rotations are involved, e.g., in the case of purely axial stretching. Such parameters, which influence the robustness of the iteration process, will be denoted as *critical parameters*. They may be the same parameters that are responsible for locking. As shown in Section 3.5.4, material parameters, such as the Poisson's ratio, can be critical parameters, too. For nearly incompressible problems, small deviations from volume preserving states yield high deviations in stress due to the activation of non-isochoric deformation modes.

Remark 3.6. *The EAS approach of Section 3.2.2.2 is not applied for this model problem, since it is identical with the displacement-based approach* (3.19). *Any enhancement applied to a bar problem would vanish because of the constant strain assumed within the bar and the patch test requirements on the enhanced field. The equivalence of EAS and displacement-based approach implies especially, that the EAS method suffers from the same poor extrapolation of the axial force as the displacement-based approach.*

Remark 3.7. *In the context of asymptotic stability analyses the authors of [44] denoted the phenomena as "extrapolation locking" (or "bifurcation locking"), since the issues described above can affect the discrete solutions in the sense that critical loads are overestimated. This denomination will not be used here, since for standard boundary value problems the aforementioned issues do, in contrast to "classical" locking phenomena, not affect the converged solution.*

3.3.2. Modified displacement-based NR-schemes

To improve the convergence properties of the displacement-based formulation (3.19), without using mixed methods, the principal idea is to modify the update of the stress (normal force, respectively) that is used for the geometric tangent.

A simple strategy for a modified NR scheme is presented by Kuo-Mo [72]. The idea is to use the converged stress from the previous load increment for the entire iteration process of the current load increment. This is denoted as *previous stress* (PS) method in the following. The influence of inaccurately extrapolated stress is avoided and the method yields a robust iteration process, independent of critical parameters. However, quadratic convergence of the method close to the solution is destroyed. This effect is particularly pronounced for large load increments, since the modified tangent is not close to the consistent one. Thus, in many situations a high number of iterations is necessary which can make the method uneconomical.

This drawback is overcome by another approach recently proposed by Magisano et al. [88]. Although for this simple model problem their approach coincides with the AS formulation, it will be used to demonstrate how the method works. Its key idea is to take the AS approach (3.20) and conduct static condensation of the independent stress increment

$$\Delta N = EA\,(\varepsilon - \hat{\varepsilon} + (1+u)\Delta u + w\Delta w) = \hat{N} - N + \Delta\hat{N}, \tag{3.23}$$

where $\Delta\hat{N} = EA((1+u)\Delta u + w\Delta w) = EA\Delta\varepsilon$. Then, for iteration (j), the update of the recovered normal force $N^{(j+1)} = N^{(j)} + \Delta N^{(j)}$ can be rewritten solely in terms of constitutive stress quantities as

$$N^{(j+1)} = \hat{N}^{(j)} + \Delta\hat{N}^{(j)} = EA(\varepsilon^{(j)} + \Delta\varepsilon^{(j)}), \tag{3.24}$$

which differs from the standard displacement-based stress update $\hat{N}^{(j+1)} = EA\varepsilon(u^{(j+1)}, w^{(j+1)})$. The constitutive force increment $\Delta\hat{N}^{(j)}$ can be interpreted as a linearized constitutive extrapolated force. It provides a reduced extrapolation error and facilitates the improved robustness. For the given model problem, the reduced linearized form, obtained after static condensation, is equivalent to the displacement-based version (3.19) with the exception of $\hat{N}$ in the tangent being replaced by (3.24). In analogy to (3.31) in the next section, we may rewrite (3.24) in terms of the current displacement degrees of freedom $\mathbf{d}^{(j)}$, which yields

$$N_{\text{MIP/AS}}^{(j+1)} := N^{(j+1)} = EA\varepsilon(\mathbf{d}^{(j)}) + EA\mathbf{B}_{\mathrm{d}}^{(j)}\Delta\mathbf{d}^{(j)}, \tag{3.25}$$

where $\mathbf{B}_{\mathrm{d}}^{(j)} = \begin{bmatrix} 1 + u^{(j)} & w^{(j)} \end{bmatrix}$ is the incremental strain-displacement operator matrix of the bar element. The authors of [88] denoted this approach as *mixed integration point method* (MIP), since its derivation for shells and solids requires an assumed stress formulation on Gauss point level, as presented in Section 3.4. For the simple model problem studied in this section, MIP and AS yield identical results, which is why they are used synonymously in (3.25). However, this is not the case when nonlinear constitutive relations or more advanced structural elements are considered, which will be emphasized in detail in the subsequent sections.

It should also be mentioned that the displacement-based formulation and the AS formulation yield the same equilibrium solution for this model problem. Thus, the MIP approach, which is basically a reformulation of the AS method, converges to this result as well. This means that the tangent close to the equilibrium is the same for both methods and thus quadratic convergence is maintained.

3.3.3. Numerical investigation

Next, a numerical study of the introduced model problem illustrates the differences between the displacement-based (U), previous stress (PS), assumed stress (AS) and mixed integration point (MIP) method. For the fixed values $c = 0.25$ and $k = 10$ an applied load $\lambda = 4.5$ is still below the critical load factor for $EA \geq 10^2$, see Figure 3.1. Figure 3.2 shows the number of total NR iterations for different values of EA, needed for convergence if the load is applied within one load increment (left) and five load increments (right).

It can be observed that the robustness[6] of the displacement-based formulation depends on the axial stiffness EA (critical parameter). For increasing values of EA, the required number of iterations to fulfill the convergence criterion increases significantly. The results of the PS method are independent of the critical parameter EA, but the required number of iterations is high due to impaired convergence properties. In sharp contrast to that, the results of both the MIP and the AS method are *independent* of EA and significantly more efficient than the PS method.

[6] Throughout the remainder of this work the term robustness is used to describe the insensitivity of the NR procedure to the values of critical parameters.

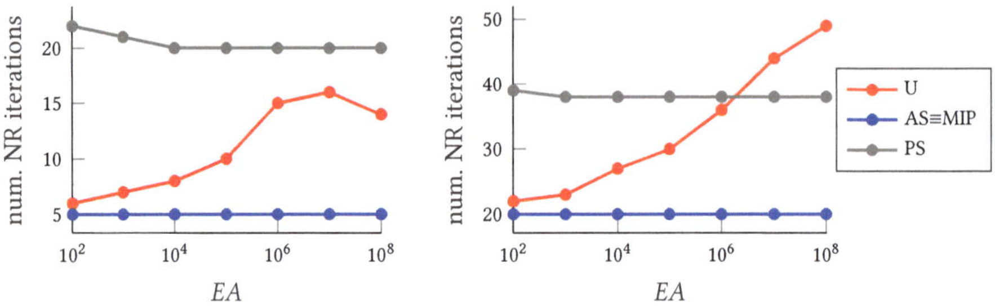

Figure 3.2.: Model problem. Cumulative number of NR iterations for $\lambda = 4.5$ in dependence of the axial stiffness EA with convergence criterion $||\Delta\mathbf{d}|| < 10^{-7}$ (update of displacement DOFs). 1 load increment (left); 5 load increments (right).

Table 3.1.: Model problem. Normal force (until converged) which are used in each iteration step (j) to update the geometric tangent, 1 load increment $\lambda = 4.5$.

j	$\hat{N}^{(j)} = \hat{N}(u^{(j)}, w^{(j)})$			$N^{(j)}_{\text{MIP/AS}} = \hat{N}^{(j-1)} + \Delta\hat{N}^{(j-1)}$		
EA=	10^2	10^3	10^4	10^2	10^3	10^4
0	0	0	0	0	0	0
1	−3.766	1.838	58.780	−4.500	−4.500	−4.500
2	−4.398	−3.523	−3.616	−4.819	−4.598	−4.578
3	−4.839	−3.588	23.820	−4.867	−4.624	−4.603
4	−4.868	−4.584	−4.432	−4.868	−4.624	−4.603
5	−4.868	−4.623	−3.811	–	–	–
6	–	−4.624	−4.602	–	–	–
7	–	–	−4.603	–	–	–

These results are confirmed in Table 3.1, where the normal forces which occur within the iteration process are shown. At the beginning of the iteration process strongly deviating values, w.r.t. the equilibrated force, are obtained for the displacement-based formulation, especially for large values of *EA*. An axial stiffness of $EA = 10^4$ for instance, even yields an artificially high constitutive tensile force in the first iteration step, emanating from the linearized (not yet correct) deformed configuration. In contrast to this, the values provided by the MIP and AS method are close to the converged one from the beginning, independent of the value of *EA*.

Figure 3.3 visualizes the spaces of the normal forces spanned by $u - w$, as well as the values for the first 4 iterations (dots) for the case of $EA = 10^4$ and a single load increment. The black curve corresponds to the equilibrated normal force for $4.5 \geq \lambda \geq 0$. The space of the displacement-based force (in red, left) is defined by (3.18) and remains the same quadratic surface for all iterations. On the other hand the space of the MIP/AS force (in blue, right) is defined by (3.25) and consists of different planes for each iteration step (j). Here, a hollow dot indicates the force $N^{(j)}_{\text{MIP/AS}}$, which is extrapolated at the current plane from the displacement-based force $\hat{N}^{(j-1)}$ (full dot) as shown in (3.25). Note especially, that the

MIP/AS has visually converged to the black dot (equilibrium), while the displacement-based formulation is still far off that point. This shows the high robustness of the MIP/AS formulation and is in line with the results in Table 3.1. The essence of Figure 3.3 is, that the linear extrapolated force in every step of the MIP/AS approach is a better guess than the nonlinear extrapolation of the displacement-based formulation.

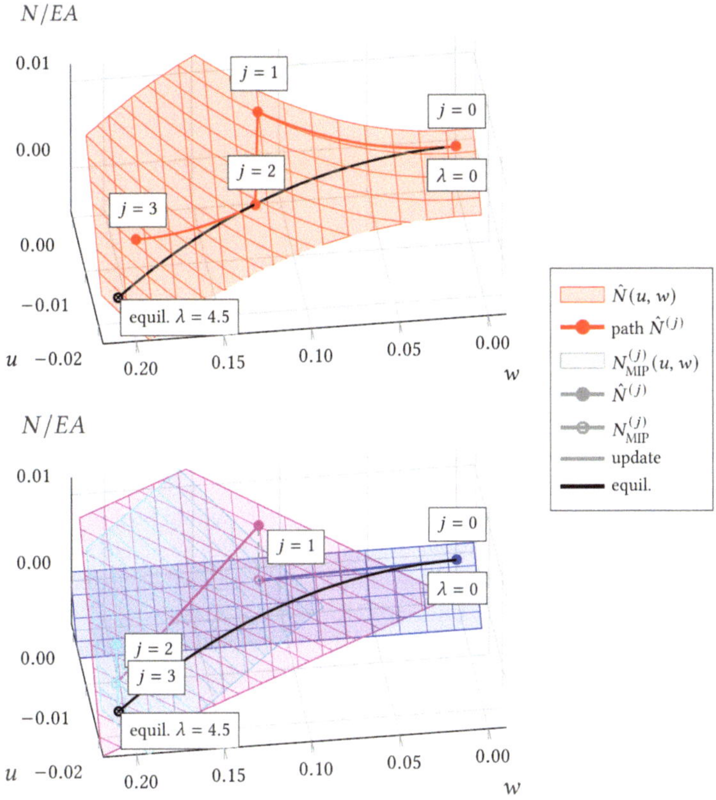

Figure 3.3.: Model problem. Visualization of the normal force within the first 4 NR iterations for one load increment, $\lambda = 4.5$ and $EA = 10^4$. Displacement-based method (top) and MIP/AS approach (bottom).

3.4. MIP method for displacement-based finite elements and general material models

In this section, we first show the extension of the MIP concept to displacement-based solid finite elements, following Magisano et al. [88]. Next, it is applied to general nonlinear material models, thus overcoming one of the major limitations of the MIP concept. The discrete weak form for AS elements (3.15) serves as basis for the MIP method. Usually, the stress in (3.15) are approximated with elementwise continuous ansatz functions. In contrast to this, the key idea of the MIP method for shell and solid problems is to choose pointwise stress approximations. Therefore, independent stresses $\mathbf{S}_g$ are introduced at every Gauss quadrature point g (see Magisano et al. [88]). Using Gauss quadrature with weights w_g, the discrete weak form on element level can be written as

$$G_{\mathbb{m},\varphi} = \sum_{g=1}^{n} \delta\boldsymbol{\varphi}^e \cdot w_g \left[\left(\mathbf{B}^{\varphi}_{\mathrm{d},g}\right)^{\mathrm{T}} \mathbf{S}_g \right], \tag{3.26a}$$

$$G_{\mathbb{m},S} = \sum_{g=1}^{n} \delta\mathbf{S}_g \cdot w_g \left[\mathbf{E}_{\varphi,g} - \hat{\mathbf{E}}_g\right] = 0, \tag{3.26b}$$

where n is the number of Gauss points and $\mathbb{m}$ denotes the weak form of the MIP method. The nodal operator matrix $\mathbf{B}^{\varphi}_{\mathrm{d},g}$ is the standard strain-displacement matrix for displacement-based elements, see Section 3.2.2.1. The constitutive strain $\hat{\mathbf{E}}_g$ is given by (3.16), while $\mathbf{E}_{\varphi,g}$ is computed via (3.3) from the deformations $\boldsymbol{\varphi}$ alone. Due to the pointwise definition of the independent stress, (3.26b) yields n independent equations (one at each Gauss point) of the form

$$\hat{\mathbf{E}}_g = \mathbf{E}_{\varphi,g}, \tag{3.27}$$

since $\delta\mathbf{S}_g$ is arbitrary. Inserting this relation into (3.26a) would lead to the pure displacement-based formulation (3.8). However, herein system (3.26) is used to increase robustness of displacement-based elements in the Newton-Raphson scheme, as demonstrated by Magisano et al. [88]. To that end, the linearized form of (3.26) is needed. After considering that the variations $\delta\boldsymbol{\varphi}^e$ and $\delta\mathbf{S}_g$ are arbitrary this is given by

$$\begin{bmatrix} \sum_{g=1}^{n} w_g \mathbf{G}^{\varphi\varphi}_{\mathrm{d},g}(\mathbf{S}_g) & w_1 \left(\mathbf{B}^{\varphi}_{\mathrm{d},1}\right)^{\mathrm{T}} & \cdots & w_n \left(\mathbf{B}^{\varphi}_{\mathrm{d},n}\right)^{\mathrm{T}} \\ \mathbf{B}^{\varphi}_{\mathrm{d},1} & -\hat{\mathbb{D}}_1 & & \mathbf{0} \\ \vdots & & \ddots & \\ \mathbf{B}^{\varphi}_{\mathrm{d},n} & \mathbf{0} & & -\hat{\mathbb{D}}_n \end{bmatrix} \begin{bmatrix} \Delta\boldsymbol{\varphi}^e \\ \Delta\mathbf{S}_1 \\ \vdots \\ \Delta\mathbf{S}_n \end{bmatrix} = - \begin{bmatrix} \sum_{g=1}^{n} w_g \left(\mathbf{B}^{\varphi}_{\mathrm{d},g}\right)^{\mathrm{T}} \mathbf{S}_g \\ \mathbf{E}_{\varphi,1} - \hat{\mathbf{E}}_1 \\ \vdots \\ \mathbf{E}_{\varphi,n} - \hat{\mathbf{E}}_n \end{bmatrix}, \tag{3.28}$$

with the tangential compliance $\hat{\mathbb{D}}_g$, defined in (3.17). The geometric part of the tangent $\mathbf{G}^{\varphi\varphi}_{\mathrm{d},g}$ has already been introduced in (3.9c), but here it is evaluated from the independent stress $\mathbf{S}_g$, which in general differ from $\hat{\mathbf{S}}_{\varphi,g}$. This is the key to the MIP method derived below.

In the next step, static condensation of the independent stress $\mathbf{S}_g$ is performed in analogy to the derivations in Section 3.3. The last n rows of (3.28) yield

$$\Delta\mathbf{S}_g = \hat{\mathbb{C}}_g \mathbf{B}^{\varphi}_{\mathrm{d},g} \Delta\boldsymbol{\varphi}^e + \hat{\mathbb{C}}_g \mathbf{E}_{\varphi,g} - \hat{\mathbb{C}}_g \hat{\mathbf{E}}_g, \tag{3.29}$$

with $\hat{\mathbb{C}}_g = \hat{\mathbb{C}}(\hat{\mathbf{E}}_g) = \hat{\mathbb{D}}_g^{-1}$. Inserting this result into the first row of (3.28) leads to

$$\sum_{g=1}^{n} w_g \left[\left(\mathbf{B}_{\mathrm{d},g}^{\varphi}\right)^{\mathrm{T}} \hat{\mathbb{C}}_g \mathbf{B}_{\mathrm{d},g}^{\varphi} + \mathbf{G}_{\mathrm{d},g}^{\varphi\varphi}(\mathbf{S}_g) \right] \Delta\boldsymbol{\varphi}^e = -\sum_{g=1}^{n} w_g \left(\mathbf{B}_{\mathrm{d},g}^{\varphi}\right)^{\mathrm{T}} \left[\mathbf{S}_g + \hat{\mathbb{C}}_g \left(\mathbf{E}_{\varphi,g} - \hat{\mathbf{E}}_g\right) \right], \tag{3.30}$$

which is rather similar to (3.8). It includes, however, a major disadvantage, which is the necessity of computing $\hat{\mathbf{E}}_g$ from the inverse stress-strain relation (see, e.g., 3.A.2.2). For linear stress-strain relations, i.e., St. Venant-Kirchhoff material, (3.29) and (3.30) can be greatly simplified. In this case, $\hat{\mathbb{C}}_g = \hat{\mathbb{C}}_{\varphi,g}$ and $\hat{\mathbb{C}}_g(\mathbf{E}_{\varphi,g} - \hat{\mathbf{E}}_g) = \hat{\mathbf{S}}_{\varphi,g} - \mathbf{S}_g$ hold, which yields

$$\mathbf{S}_g^{(j+1)} = \mathbf{S}_g^{(j)} + \Delta\mathbf{S}_g^{(j)} = \left[\hat{\mathbf{S}}_{\varphi,g} + \hat{\mathbb{C}}_{\varphi,g} \mathbf{B}_{\mathrm{d},g}^{\varphi} \Delta\boldsymbol{\varphi}^e \right]^{(j)} \tag{3.31}$$

for the update procedure of the independent stress at every Gauss point within iteration (j) of the global Newton-Raphson scheme. Moreover, (3.30) reduces accordingly to

$$\sum_{g=1}^{n} w_g \left[\left(\mathbf{B}_{\mathrm{d},g}^{\varphi}\right)^{\mathrm{T}} \hat{\mathbb{C}}_{\varphi,g} \mathbf{B}_{\mathrm{d},g}^{\varphi} + \mathbf{G}_{\mathrm{d},g}^{\varphi\varphi}(\mathbf{S}_g) \right] \Delta\boldsymbol{\varphi}^e = -\sum_{g=1}^{n} w_g \left(\mathbf{B}_{\mathrm{d},g}^{\varphi}\right)^{\mathrm{T}} \hat{\mathbf{S}}_{\varphi,g}, \tag{3.32}$$

which is identical to (3.8), with the exception of $\mathbf{S}_g$ being used for computing the geometric tangent instead of $\hat{\mathbf{S}}_{\varphi,g}$. The right-hand side remains unchanged, which means that the equilibrium solutions of (3.32) and (3.8) coincide. Only the tangent is modified, which ultimately leads to improved robustness in the Newton-Raphson scheme, as demonstrated in the numerical simulations in Section 3.6. In contrast to the simple example from Section 3.3, the MIP method for solid finite elements is not equivalent to the standard HR formulation due to the pointwise, instead of elementwise, approximations of the stress.

Equations (3.31) and (3.32) are the core of the MIP method proposed by Magisano et al. [88], who are taking linear material behavior into account. In case of *general material models*, we assume that (3.31) and (3.32) still hold, at least approximately. This implies that assumptions

$$\hat{\mathbb{C}}_g \approx \hat{\mathbb{C}}_{\varphi,g} \qquad \text{and} \qquad \hat{\mathbb{C}}_g(\mathbf{E}_{\varphi,g} - \hat{\mathbf{E}}_g) \approx \hat{\mathbf{S}}_{\varphi,g} - \mathbf{S}_g \tag{3.33}$$

are made. As the right hand side in (3.32) does not require inversion of the stress-strain relation, it is easily applied to general material models. Assumptions (3.33) are always fulfilled in the first NR iteration of any load step since in equilibrium (3.27) holds and thus also $\mathbf{S}_g = \hat{\mathbf{S}}_{\varphi,g}$. Furthermore, the assumptions are true for NR iterations close to equilibrium (i.e., in the range of quadratic convergence) since then $||\Delta\boldsymbol{\varphi}^{(j)}|| \ll 1$ and we get from (3.31) $\mathbf{S}_g \approx \hat{\mathbf{S}}_{\varphi,g}$ respectively $\mathbf{E}_{\varphi,g} \approx \hat{\mathbf{E}}_g$. A final case, for which (3.33) holds, are states with small strain. Under that limitation reasonable material models are almost identical to the St. Venant-Kirchhoff model in order to ensure consistency with linear theory. Thus, (3.33) is fulfilled since it is identically fulfilled for the St. Venant-Kirchhoff model.

The implications of assumption (3.33) are studied in the numerical simulations presented in Section 3.6. There, it is shown that the proposed MIP method (3.32) works best as long as the problem is limited to small strain. As soon as large strain occur, the method becomes less beneficial but still provides advantageous behavior in many cases.

Remark 3.8. *The pointwise stress ansatz chosen in* (3.28) *and its condensed version* (3.30) *increase robustness, but do not allow the design of locking-free finite elements due to the principle of limitation by Fraeijs de Veubeke [42] (see Zienkiewicz et al. [162] Ch. 10.2).*

3.5. Further extensions and details for the MIP method

3.5.1. EAS finite elements

In this section, the MIP method is extended to EAS elements in order to overcome one of the major drawbacks of this highly popular element class, which is the lack of robustness in the Newton-Raphson scheme (see [65, 145]).

In a first step, the weak forms of the EAS method (3.11) and the AS method (3.15) are combined by replacing the constitutive stress $\hat{\mathrm{S}}_\alpha$ in (3.11) with the independent stress S_β from (3.15). Additionally, (3.15b) is utilized to enforce the constitutive law. The resulting weak form of the three-field formulation is given by

$$G_{\tilde{\mathbb{m}},\varphi} = \int_{\Omega^e} \mathrm{S}_\beta : \delta_{\boldsymbol{\varphi}} \mathrm{E}_\alpha \, \mathrm{d}V, \tag{3.34a}$$

$$G_{\tilde{\mathbb{m}},\alpha} = \int_{\Omega^e} \mathrm{S}_\beta : \delta_{\boldsymbol{\alpha}} \mathrm{E}_\alpha \, \mathrm{d}V = 0, \tag{3.34b}$$

$$G_{\tilde{\mathbb{m}},S} = \int_{\Omega^e} \delta\mathrm{S}_\beta : (\mathrm{E}_\alpha - \hat{\mathrm{E}}_\beta) \, \mathrm{d}V = 0 \tag{3.34c}$$

and marked with index $\tilde{\mathbb{m}}$. In a second step, we approximate the independent stress S_β, as in (3.26), in a pointwise manner at the Gauss points and employ Gauss quadrature to get

$$\begin{bmatrix} \sum_{g=1}^{n} w_g \mathbf{G}^{\varphi\varphi}_{\mathbb{e},g}(\mathrm{S}_g) & \sum_{g=1}^{n} w_g \mathbf{G}^{\varphi\alpha}_{\mathbb{e},g}(\mathrm{S}_g) & w_1 \left(\mathbf{B}^{\varphi}_{\mathbb{e},1}\right)^{\mathrm{T}} & \cdots & w_n \left(\mathbf{B}^{\varphi}_{\mathbb{e},n}\right)^{\mathrm{T}} \\ \sum_{g=1}^{n} w_g \mathbf{G}^{\alpha\varphi}_{\mathbb{e},g}(\mathrm{S}_g) & \sum_{g=1}^{n} w_g \mathbf{G}^{\alpha\alpha}_{\mathbb{e},g}(\mathrm{S}_g) & w_1 \left(\mathbf{B}^{\alpha}_{\mathbb{e},1}\right)^{\mathrm{T}} & \cdots & w_n \left(\mathbf{B}^{\alpha}_{\mathbb{e},n}\right)^{\mathrm{T}} \\ \mathbf{B}^{\varphi}_{\mathbb{e},1} & \mathbf{B}^{\alpha}_{\mathbb{e},1} & -\hat{\mathbb{D}}_1 & & \mathbf{0} \\ \vdots & \vdots & & \ddots & \\ \mathbf{B}^{\varphi}_{\mathbb{e},n} & \mathbf{B}^{\alpha}_{\mathbb{e},n} & \mathbf{0} & & -\hat{\mathbb{D}}_n \end{bmatrix} \begin{bmatrix} \Delta\boldsymbol{\varphi}^e \\ \Delta\boldsymbol{\alpha}^e \\ \Delta\mathrm{S}_1 \\ \vdots \\ \Delta\mathrm{S}_n \end{bmatrix} = - \begin{bmatrix} \sum_{g=1}^{n} w_g \left(\mathbf{B}^{\varphi}_{\mathbb{d},g}\right)^{\mathrm{T}} \mathrm{S}_g \\ \sum_{g=1}^{n} w_g \left(\mathbf{B}^{\alpha}_{\mathbb{d},g}\right)^{\mathrm{T}} \mathrm{S}_g \\ \mathrm{E}_{\alpha,1} - \hat{\mathrm{E}}_1 \\ \vdots \\ \mathrm{E}_{\alpha,n} - \hat{\mathrm{E}}_n \end{bmatrix} \tag{3.35}$$

which is the linearized form of (3.34) on element level. The static condensation procedure, as presented in Section 3.4, leads in this case to

$$\mathbf{S}_g^{(j+1)} = \left[\hat{\mathbf{S}}_{\alpha,g} + \hat{\mathbb{C}}_{\alpha,g} \mathbf{B}_{\mathrm{e},g}^{\varphi} \Delta \boldsymbol{\varphi}^e + \hat{\mathbb{C}}_{\alpha,g} \mathbf{B}_{\mathrm{e},g}^{\alpha} \Delta \boldsymbol{\alpha}^e \right]^{(j)} \tag{3.36}$$

for the update of the independent stress at the Gauss points in step (j) of the Newton-Raphson scheme. This finally yields the linearized weak form of the MIP method for EAS elements given by

$$\sum_{g=1}^{n} w_g \begin{bmatrix} \mathbf{M}_{\mathrm{e},g}^{\varphi\varphi} + \mathbf{G}_{\mathrm{e},g}^{\varphi\varphi}(\mathbf{S}_g) & \mathbf{M}_{\mathrm{e},g}^{\varphi\alpha} + \mathbf{G}_{\mathrm{e},g}^{\varphi\alpha}(\mathbf{S}_g) \\ \mathbf{M}_{\mathrm{e},g}^{\alpha\varphi} + \mathbf{G}_{\mathrm{e},g}^{\alpha\varphi}(\mathbf{S}_g) & \mathbf{M}_{\mathrm{e},g}^{\alpha\alpha} + \mathbf{G}_{\mathrm{e},g}^{\alpha\alpha}(\mathbf{S}_g) \end{bmatrix} \begin{bmatrix} \Delta \boldsymbol{\varphi}^e \\ \Delta \boldsymbol{\alpha}^e \end{bmatrix} = - \sum_{g=1}^{n} w_g \begin{bmatrix} \mathbf{R}_{\mathrm{e},g}^{\varphi} \\ \mathbf{R}_{\mathrm{e},g}^{\alpha} \end{bmatrix}, \tag{3.37}$$

where the only difference compared to (3.13) is the use of the independent stress $\mathbf{S}_g$ instead of constitutive stress $\hat{\mathbf{S}}_{\alpha,g}$ to compute the tangent. In analogy to Section 3.4, the right hand side (i.e., the residual) is not altered. Thus, the converged solutions of (3.37) and (3.13) are identical. However, (3.37) improves robustness of the EAS method in NR iterations as shown in Section 3.6. Static condensation of the enhanced degrees of freedom $\boldsymbol{\alpha}^e$ can be performed on element level as usual. The extension of the MIP method for EAS elements to general material models is straightforward and analogous to Section 3.4. This implies especially that assumptions (3.33) are made in order to avoid the need for an inverse material law.

3.5.2. Spatial formulation of the MIP method

For many material laws, that are formulated on the basis of spatial measures, finite element formulations in the current configuration are advantageous due to higher numerical efficiency and simpler implementation (see, e.g., [154]). Because of these benefits, in this section, the MIP method is transformed from the reference configuration (Sections 3.4 and 3.5.1) to its spatial form.

A push forward of the AS weak form (3.15), which is the basis for the displacement-based MIP method (see Section 3.4), yields

$$G_{\mathfrak{s},\varphi} = \int_{\Omega^e} \mathbf{S}_\beta : \delta_\varphi \mathbf{E}_\varphi \, \mathrm{d}V \qquad = \int_{\Omega^e} \boldsymbol{\tau}_\beta : \delta_\varphi \mathbf{e}_\varphi \, \mathrm{d}V, \tag{3.38a}$$

$$G_{\mathfrak{s},\beta} = \int_{\Omega^e} \delta_\beta \mathbf{S}_\beta : (\mathbf{E}_\varphi - \hat{\mathbf{E}}_\beta) \, \mathrm{d}V = \int_{\Omega^e} \delta_\beta \boldsymbol{\tau}_\beta : (\mathbf{e}_\varphi - \hat{\mathbf{e}}_\beta) \, \mathrm{d}V = 0, \tag{3.38b}$$

where $\boldsymbol{\tau}_\beta := \mathbf{F}_\varphi \mathbf{S}_\beta \mathbf{F}_\varphi{}^{\mathrm{T}}$ is the (independent) *Kirchhoff* stress tensor. Furthermore, $\mathbf{e}_\varphi = \frac{1}{2}(\mathbf{I} - \mathbf{F}_\varphi^{-\mathrm{T}} \mathbf{F}_\varphi^{-1})$ and $\hat{\mathbf{e}}_\beta = \mathbf{F}_\varphi^{-\mathrm{T}} \hat{\mathbf{E}}_\beta \mathbf{F}_\varphi^{-1}$ denote the displacement-based and constitutive *Almansi* strain tensor, respectively. However, in the approximated regime two ways to introduce the pointwise stress at the Gauss points emerge. One could, as in previous sections, approximate the material second Piola-Kirchhoff stresses $\mathbf{S}_g$ or instead define new independent stresses $\boldsymbol{\tau}_g := \mathbf{F}_\varphi \mathbf{S}_g \mathbf{F}_\varphi{}^{\mathrm{T}}$ and use that as unknowns. Our numerical experiments have shown that the latter approach is inferior to the pointwise approximation of S, which is why only this approach is considered here.

Proceeding similarly with the linearized form of (3.38) and performing the steps presented in Section 3.4 ultimately yields the MIP weak form and the stress update formula in the current configuration given by

$$\sum_{g=1}^{n} w_g \left[\left(\mathbf{b}^{\varphi}_{\mathrm{d},g}\right)^{\mathrm{T}} \hat{\mathbb{c}}_{\varphi,g} \mathbf{b}^{\varphi}_{\mathrm{d},g} + \mathbf{g}^{\varphi\varphi}_{\mathrm{d},g}(\mathbf{F}_\varphi \mathbf{S}_g \mathbf{F}^{\mathrm{T}}_\varphi) \right] \Delta\boldsymbol{\varphi}^e = - \sum_{g=1}^{n} w_g \left(\mathbf{b}^{\varphi}_{\mathrm{d},g}\right)^{\mathrm{T}} \hat{\boldsymbol{\tau}}_{\varphi,g}, \tag{3.39}$$

$$\mathbf{S}^{(j+1)}_g = \mathbb{F}\left(\left(\mathbf{F}^{(j)}_\varphi\right)^{-1}\right) \left[\hat{\boldsymbol{\tau}}_{\varphi,g} + \hat{\mathbb{c}}_{\varphi,g} \mathbf{b}^{\varphi}_{\mathrm{d},g} \Delta\boldsymbol{\varphi}^e \right]^{(j)}, \tag{3.40}$$

where $\mathbf{b}^{\varphi}_{\mathrm{d},g}$ are the spatial nodal operator matrices and $\hat{\mathbb{c}}_\varphi$ denotes the spatial material tangent, which is given in index notation by $\hat{\mathbb{c}}_{abcd} = F_{aA} F_{bB} F_{cC} F_{dD} \hat{\mathbb{C}}_{ABCD}$. Moreover,

$$\mathbb{F}(\mathbf{F}) = \begin{bmatrix} F^2_{11} & F^2_{12} & F^2_{13} & 2F_{11}F_{12} & 2F_{12}F_{13} & 2F_{11}F_{13} \\ F^2_{21} & F^2_{22} & F^2_{23} & 2F_{21}F_{22} & 2F_{22}F_{23} & 2F_{21}F_{23} \\ F^2_{31} & F^2_{32} & F^2_{33} & 2F_{31}F_{32} & 2F_{32}F_{33} & 2F_{31}F_{33} \\ F_{11}F_{21} & F_{12}F_{22} & F_{13}F_{23} & F_{11}F_{22} + F_{12}F_{21} & F_{12}F_{23} + F_{13}F_{22} & F_{11}F_{23} + F_{13}F_{21} \\ F_{21}F_{31} & F_{22}F_{32} & F_{23}F_{33} & F_{21}F_{32} + F_{22}F_{31} & F_{22}F_{33} + F_{23}F_{32} & F_{21}F_{33} + F_{23}F_{31} \\ F_{11}F_{31} & F_{12}F_{32} & F_{13}F_{33} & F_{11}F_{32} + F_{12}F_{31} & F_{12}F_{33} + F_{13}F_{32} & F_{11}F_{33} + F_{13}F_{31} \end{bmatrix} \tag{3.41}$$

performs the transformation $\boldsymbol{\tau}_g = \mathbf{F}_\varphi \mathbf{S}_g \mathbf{F}_\varphi{}^{\mathrm{T}}$ in vector-matrix form. All in all, the only necessary changes to get from material to spatial MIP form is exchanging all material variables in (3.31) and (3.32) with their spatial counterparts and perform the transformation using (3.41). We emphasize that (3.39) and (3.40) are equivalent to (3.32) and (3.31), respectively.

A spatial version is also easily obtained for the MIP method for EAS elements. The same steps lead to a similar result, where all quantities in the reference configuration in (3.37) and (3.36) have to be replaced by corresponding spatial variables. For the sake of brevity, we omit corresponding derivations.

3.5.3. Implementation

The implementation of the MIP method described in the previous two sections is relatively simple. Only a few changes have to be made in existing codes. A possible global code structure in pseudo-code format is given in Algorithm 3.1 for the EAS element with MIP extension. The corresponding element routine is shown in Algorithm 3.2. In those algorithms, all changes and additional operations necessary for the MIP method are marked in red.

The Algorithms 3.1 and 3.2 are generally applicable regardless of material model or solution routine. For instance, they can be applied for elasto-plastic simulations even if a line-search modified Newton-Raphson scheme is used (see Section 3.6.6).

Algorithm 3.1.: MIP-modified FE code for EAS element. Necessary modifications/additions to standard Newton-Raphson procedure marked in red.

```
for l = 1, ..., n_steps do                                          ▹ loop over all load steps l
    j = 0
    while not converged do                                          ▹ Newton-Raphson loop
        element routine (see Algorithm 3.2)                         ▹ element routine
                                                                    ▹ global solver
        assembly
        if ||R|| < tol then
            converged, next load step l
        else
            solution of linear equation K^(j) Δφ^(j) = −R^(j)
            deformations φ^(j+1) = φ^(j) + Δφ^(j)                   ▹ update
            enhanced parameters α^(j+1) (static condensation)
            MIP stress S_g^(j+1) = [Ŝ_α,g + Ĉ_α,g B^φ_e,g Δφ^e + Ĉ_α,g B^α_e,g Δα^e]^(j)
        end if
        j = j + 1
    end while
end for
```

Algorithm 3.2.: MIP modified element routine for EAS element. Necessary modifications/additions to standard Newton-Raphson procedure marked in red.

```
for e = 1, ..., n_el do                                             ▹ loop over all elements e
    for g = 1, ..., n_gp do                                         ▹ loop over all Gauss points g
        Ŝ_α,g = Ŝ(E_α,g, Ξ_g)
        if j = 0 then                                               ▹ j is the Newton-counter (see Algorithm 3.1)
            S_g^(j) = Ŝ_α,g
        else
            S_g^(j) = S_g^(j)
        end if
        compute residual and tangent:
          R^φ_e,g(Ŝ_α,g); R^α_e,g(Ŝ_α,g);
          M^φφ_e,g + G^φφ_e,g(S_g); M^φα_e,g + G^φα_e,g(S_g); M^αφ_e,g + G^αφ_e,g(S_g); M^αα_e,g + G^αα_e,g(S_g);
        save info for update of S_g: Ŝ_α,g; (Ĉ_α,g B^φ_e,g); (Ĉ_α,g B^α_e,g);
    end for
    static condensation of α^e
end for
```

3.5.4. One-element test

A one-element test is presented in this section to demonstrate the effects of the MIP method on the stress and convergence properties of solid elements. In analogy to the simple example in Section 3.3, the aim is to highlight basic effects of the MIP method for solid problems and reveal differences compared to standard methods. More complex numerical investigations follow in Section 3.6.

The problem setup of the one-element test investigated here is illustrated in Figure 3.4. The free end of the cantilever beam structure is subjected to a load $F = Et^3/L^3$ and a St. Venant-Kirchhoff material model under plane strain conditions is used (see Appendix 3.A.1). The evaluations presented in the following are carried out on Gauss point level. In particular, the spatial *Kirchhoff*-stress $\boldsymbol{\tau}$ is examined at the lower left Gauss point, as indicated in Figure 3.4.

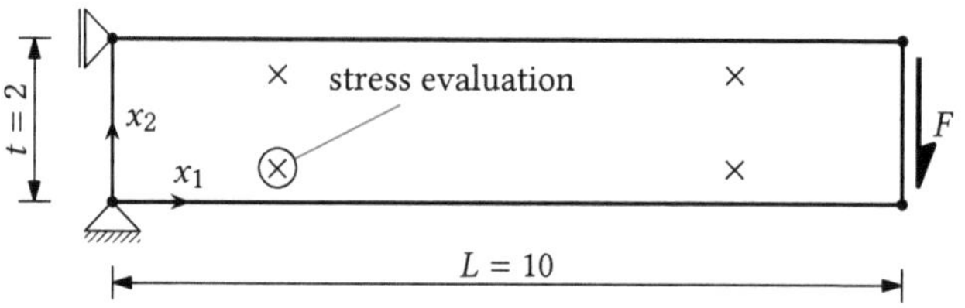

Figure 3.4.: One-element test. Problem setup with geometry, boundary conditions and Gauss points.

In a first step, the assumed stress element Q1/S5 [110] is compared to the standard EAS element Q1/E4 [131, 134] (see Section 3.6 for more detailed descriptions of the elements). These two elements are free of locking for the present undistorted case and are equivalent in linear problems (see, e.g., Bischoff et al. [21]). In the geometrically nonlinear case, the elements are no longer identical, but still can be expected to provide similar results. Figure 3.5 shows the development of the axial stress component in x_1-direction, τ_{11}, for different values of Poisson's ratio ν. The stress τ_{11} is normalized with the converged stress after all five load steps denoted by τ_{11}^{end}. The stress plotted is determined at the beginning of every iteration, which explains why the stress in the first NR iteration of a load step is the same as the previously converged result. On the horizontal axis, load steps with converged results are marked at integers with vertical lines. Every NR iteration in between is marked with an "x". For example, the third Newton iteration out of a total of five necessary iterations in the second load step is marked at $2 + 3/5 - 1 = 1.6$. For $\nu = 0$, Figure 3.5 (left) shows that the stress converges much faster for Q1/S5, which are almost converged after the first NR iteration within each load increment. This is in line with the observations of Magisano et al. [87, 88] and the 2-DOF example presented in Section 3.3. The converged results of Q1/E4 are visually identical. However, after the first NR iteration the stress is far worse, which results in an additionally required iteration in load steps 2-5. This difference becomes worse for the nearly incompressible case with $\nu = 0.499$ shown in Figure 3.5 (right). In this case, the stress partly has the wrong sign and are off by a factor of ≈ 36 after the first iteration in the first load step. This results in one to two additionally required iterations per load increment, while Q1/S5 is insensitive to the critical parameter ν.

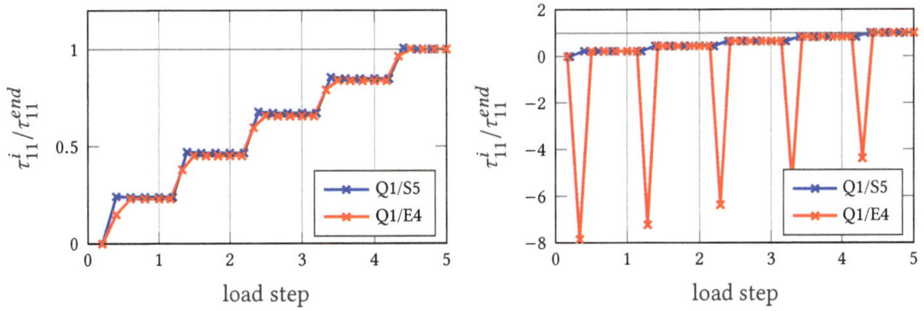

Figure 3.5.: One-element test. Comparison of Q1/S5 and Q1/E4. Development of the stress τ_{11} during the simulation with five load steps. Possion's ratio $\nu = 0$ (left) and $\nu = 0.499$ (right).

The MIP method allows to overcome this drawback which is illustrated in Figure 3.6. Therein, standard EAS element Q1/E4 is compared to Q1/E4-MIP, which uses the MIP method as described in Section 3.5.1. The MIP stress $\tau_{g,11}$ at the Gauss point (3.36) show almost identical behavior as the stress of Q1/S5, while the constitutive stress $\hat{\tau}_{11}$ (dashed line) used for the residuum are still far off in the first iterations. Using the altered $\tau_{g,11}$ for the geometric tangent enables the improved convergence behavior of elements. Furthermore the two kinds of stress converge quickly to each other and are equivalent in case of convergence. The Q1/E4-MIP element requires the same number of NR iterations as AS element Q1/S5.

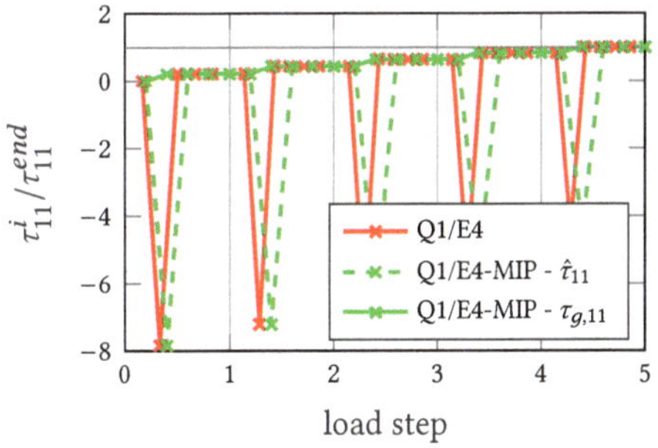

Figure 3.6.: One-element test. Comparison of Q1/E4 with Q1/E4-MIP. Development of the stress τ_{11} during the simulation with five load steps. Possion's ratio $\nu = 0.499$.

3.6. Numerical investigations

3.6.1. Overview

This section covers various numerical investigations examining the properties of the MIP method presented in Sections 3.4 and 3.5. The main concern of all investigations is robustness of the simulations, where in the context of the present work we characterize robustness by the size of applicable load steps and number of NR iterations needed to find a solution as well as the sensitivity w.r.t. to critical parameters e.g the elements' aspect ratio or Poisson's ratio. In this sense, robustness implies also efficiency, since fewer load steps with fewer NR iterations yield the desired result. It will be shown that the MIP method improves robustness in many situations but also that there exist scenarios in which little or no positive effects can be observed. Other classical topics of element technology such as locking and convergence with mesh refinement are not covered here.

All simulations are performed using 2D plane strain quadrilateral or 3D hexahedral elements. The considered element formulations are:

- **Q*p*:** Isoparametric Lagrangian displacement-based, quadrilateral with polynomial degree p. Corresponding hexahedra are denoted as **H*p***.
- **Q1/E4:** EAS element proposed by Simo and Armero [131], which employs the four Wilson-modes. In 3D the element is labeled **H1/E9** and has nine enhanced modes.
- **Q1/E4T:** EAS element employing the transposed Wilson-modes to overcome the instability of Q1/E4 under compression, as proposed by Glaser and Armero [46, 47].
- **QA1/E4T:** EAS element proposed by Pfefferkorn and Betsch [104]. Compared to Q1/E4T it uses a different quadrature rule.
- **HA1/E12T:** The 3D version of QA1/E4T. In addition to the modified quadrature rule it uses three additional enhanced modes (compared to H1/E9) and employs a special evaluation of the compatible deformation gradient.
- **Q1/S5:** Assumed stress element with a five parameter stress interpolation, as proposed for linear kinematics by Pian and Sumihara [110]. The extension to nonlinear kinematics can, e.g., be found in Viebahn et al. [145]. In 3D, the element is denoted **H1/S18** and has 18 stress modes (see, e.g., [3, 112, 145]).
- **Q*p*-MIP, Q1/E4-MIP, Q1/E4T-MIP, QA1/E4T-MIP:** Element formulations as defined above, but equipped with the MIP method as described in Sections 3.4 and 3.5. The corresponding 3D hexahedral elements are denoted as **H*p*-MIP, H1/E9-MIP, H1/E9T-MIP, HA1/E12T-MIP**.
- **Q1/E4-MIP*:** Element formulation as defined above but without considering the simplifying assumptions (3.33).

The material models used throughout this Section are a St. Venant-Kirchoff, Neo-Hookean and logarithmic strain based elasto-plastic model, which are described more detailed in Appendix 3.A.

As mentioned above, robustness is the major concern of the present work which is highly influenced by settings chosen for the NR procedure. For all simulations we use a convergence criterion based on the residual norm in the form $||\mathbf{R}|| < \epsilon_{\text{NR}}$. Note that no scaling of this norm is applied as, e.g., suggested by Belytschko et al. [15]. If not mentioned otherwise, tolerance ϵ_{NR} is set to 10^{-8} for all simulations. Failure of the NR procedure is determined if either $||\mathbf{R}|| > 10^{14}$ or more than 20 iterations are necessary to find a solution within one load step.

3.6.2. Clamped beam

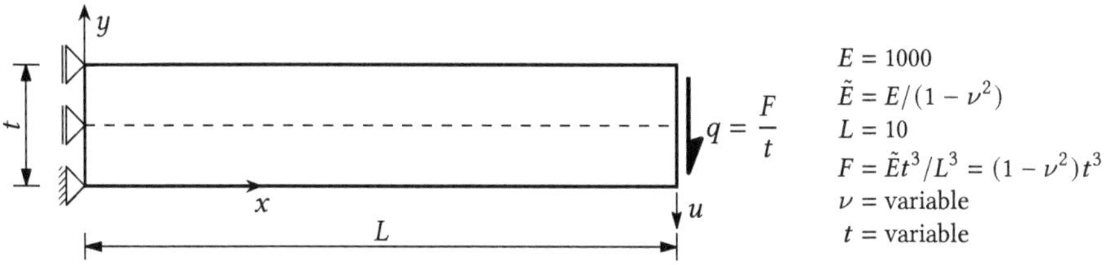

Figure 3.7.: Clamped beam. Problem setup with geometry, boundary conditions and load.

The setup of the first numerical example is shown in Figure 3.7 and consists of a rectangular block under plane strain conditions, meshed with 1×10 elements. Two elastic material models are considered for this example. The left edge is fixed horizontally and the vertical displacement is fixed only at one single point, to avoid artificial constraint stresses in y-direction. The singularity at this point has no influence for the coarse mesh considered here. On the right hand side the structure is subjected to a uniformly distributed vertical force resultant which is scaled such that the vertical tip displacement of a corresponding geometrically linear infinitely wide plate is always $u^{\text{lin}} = 4.0$. In case of $\nu = 0$ it is furthermore identical to a thin beam solution (Bernoulli). Value u^{lin} clearly deviates from the solutions for a thick beam or when a geometrically nonlinear setting is considered. However, it is a way to keep the deformation (degree of non-linearity) in a similar range for varying problem parameters, which is useful for the subsequent investigations. The displacement-based elements that will be considered in the following are the bi-quadratic Q2 element and its MIP version. They are chosen instead of the Q1 elements, since they are less susceptible to locking and thus better comparable to the other elements tested. These are the standard Q1/E4 and its MIP version and finally Q1/S5. The latter always performs most robust for this problem setup (this is not the case in general, see Section 3.6.3) and thus its number of necessary NR iterations is considered as "target" for all other elements.

First, the influence of critical *geometric* parameters, such as the beam's slenderness L/t are investigated. In order to separate different effects, Poisson's ratio is set to zero and only the thickness is varied ($t = \{1, 0.5, 0.2, 0.1, 0.05, 0.02, 0.01\}$). The load F is applied within one single load increment. The diagrams in Figure 3.8 show the cumulative number of iterations n_{NR} that are required to fulfill the convergence criterion, plotted versus the slenderness. For both materials, it can be observed that for Q2 and Q1/E4 the number of required iterations grow or the NR scheme even diverges with an increasing slenderness. This indicates that the parameter dependency of the robustness is *not* a unique phenomenon of EAS elements but rather an intrinsic behavior of constitutive-based stress updates. On the other hand, the MIP versions of both the Q2 and Q1/E4 elements exhibit robust behavior, independent of the slenderness, with only minor differences compared to Q1/S5. This test confirms the results of [88], where improved robustness was observed for the numerical analysis of thin-walled structures using solid-shells. For the case $t = 0.05$, a detailed summary of the residual norm as well as the converged tip displacement $\hat{u}$ is shown in Table 3.2. Whether MIP is applied or not does not affect the converged displacements, since the residual is not modified (see Section 3.4).

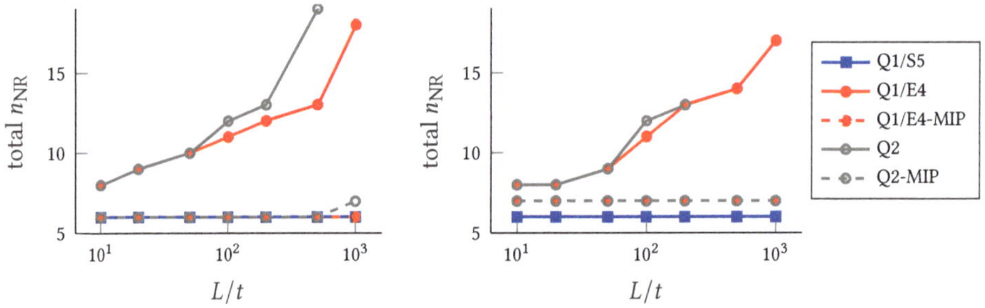

Figure 3.8.: Clamped beam. Cumulative number of NR iterations for 1 load increment in dependence of the slenderness L/t, Poisson's ratio $\nu = 0$. St. Venant-Kirchhoff (left) and Neo-Hooke (right). Here the maximum number of iterations is set to 30.

In the second parameter study, the influence of critical *material* parameters, such as the Poisson's (or bulk modulus) is investigated. In this case, the thickness of the beam is set to $t = 0.5$ and only the Poisson's ratio is varied ($\nu = \{0.0, 0.3, 0.4, 0.45, 0.49, 0.499, 0.4999\}$). Figure 3.9 shows the required cumulative number of NR iterations n_{NR} over the bulk modulus when F is applied within one (top) and five (bottom) load increment(s), respectively. For the St. Venant-Kirchhoff material a similar behavior as before is observed. All MIP elements and Q1/S5 perform robustly, whereas Q2 and Q1/E4 show, for high bulk moduli, an increasingly less robust convergence behavior. However, this outstanding performance of the MIP elements can not be re-produced completely when a Neo-Hookean material is used. Although the average response of the MIP formulations is still superior to Q2 and Q1/E4, convergence is not achieved in all cases. The reason for this bad performance can be explained by the stronger violation of assumptions (3.33). First, because larger strain occurs compared to the thin beam case and second, the non-linearity of the Neo-Hookean material directly depends on degree of incompressibility (due to term $\frac{\lambda}{2} \ln^2 J$ in (3.45)).

However, for Q1/E4-MIP*, which is the "correct" MIP formulation based on (3.35) (i.e., the simplifying assumptions (3.33) are *not* made), the same results as for Q1/S5 are obtained. Unfortunately though, the inverse stress-strain relation is needed again, which destroys the major advantage of the MIP method.

Table 3.2.: Clamped beam problem. Residual norm $||\mathbf{R}||$ during convergence process and vertical tip displacement for $\{t = 0.05,\ \nu = 0.0\}$ and St. Venant-Kirchhoff material.

n_{NR}	Q2	Q2-MIP	Q1/E4	Q1/E4-MIP	Q1S5
0	8.839e-05	8.839e-05	8.839e-05	8.839e-05	8.839e-05
1	4.316e+02	4.316e+02	5.627e+02	5.627e+02	5.627e+02
2	5.825e+01	2.449e+01	7.544e+01	2.901e+01	2.902e+01
3	2.010e+00	1.705e-01	2.570e+00	1.452e-01	1.455e-01
4	1.338e-01	3.760e-05	5.621e-02	5.997e-06	6.057e-06
5	1.743e-01	4.003e-11	6.374e-02	1.166e-11	1.038e-11
6	2.709e-02	–	6.796e-03	–	–
7	2.891e-01	–	3.654e-02	–	–
8	2.154e-03	–	2.399e-04	–	–
9	9.405e-02	–	1.760e-03	–	–
10	2.195e-05	–	3.468e-07	–	–
11	3.576e-04	–	8.692e-09	–	–
12	8.380e-10	–	–	–	–
u	3.272	3.272	3.470	3.470	3.470

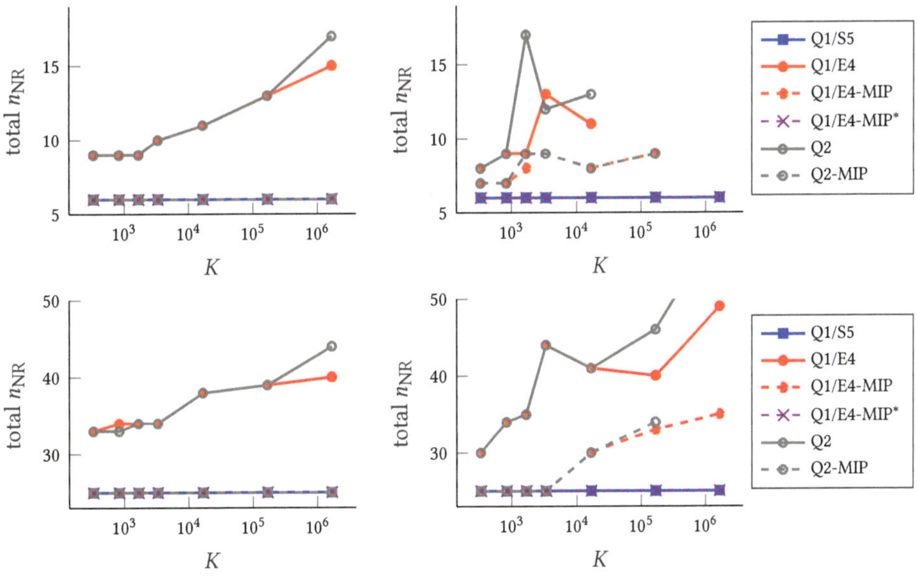

Figure 3.9.: Clamped beam. Cumulative number of NR iterations for 1 load increment (top) and 5 load increments (bottom) in dependence of the bulk modulus $K = E/(3 - 6\nu)$ for fixed thickness $t = 0.5$. St. Venant-Kirchhoff (left) and Neo-Hooke (right).

3.6.3. Elastic strip

The next example is the elastic strip test proposed by Korelc and Wriggers [69] (see also [27]), which is one of the few examples where robustness of EAS elements has been examined. Its initial square geometry ($a = 10$) and the deformed state are shown in Figure 3.10 together with the boundary conditions. The strip is loaded by prescribed displacement $u_\mathrm{p} = 10$ applied on the right edge. A regular FE-mesh with 10×10 elements is used for all simulations. As material model we consider the Neo-Hookean model described in Appendix 3.A and choose the elasticity constants to $\lambda = 24$ and $\mu = 6$. The St. Venant-Kirchhoff material is not considered for this example, since it leads to unphysical results with artificial boundary layers along the free edges as shown in Figure 3.10.

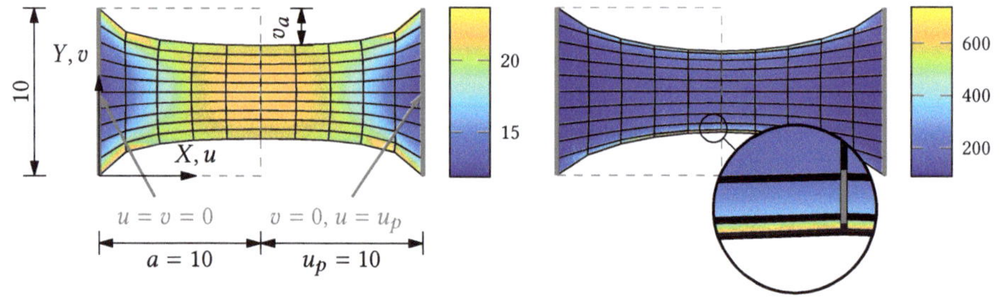

Figure 3.10.: Elastic strip. Deformed configuration with von Mises stress distribution for the Neo-Hookean (left) and St. Venant-Kirchhoff (right) material. Geometry and boundary conditions of the elastic strip example depicted on left side.

Results of the elastic strip example in 2D are shown in Table 3.3. In contrast to the clamped beam example in Section 3.6.2 there are only minor improvements in the total number of NR iterations n_NR for Q1/E4 if the MIP method is used. For other elements, namely Q1, Q2 and Q1/E4T, even more iterations are required if the MIP method is applied. This behavior follows from the very high strain occurring in this example. For these states the Neo-Hookean and St. Venant-Kirchhoff model differ noticeably which implies that assumptions (3.33) are strongly violated. Thus, the MIP approach is, for this kind of problems, a poorer approximation of an AS method and therefore less efficient. Furthermore, the MIP technique is more favorable in bending dominated problems instead of the uniaxial problem considered here. In general, it can be observed that the MIP method looses efficiency in case of large strain and general material models. However, in less extreme cases than the example considered here, there are usually still improvements due to the MIP strategy (see subsequent examples).

Very interesting are the results obtained with Q1/S5. This element employs the inversion of the stress-strain relation of the Neo-Hookean material model presented in Appendix 3.A.2.2. While it works well and converges fast for smaller displacements, failure of the simulations can be observed at $u_\mathrm{p} \approx 5.36$. At that level of deformation the inverse stress-strain relation looses its uniqueness (see Section 3.A.2.2 Remark 3.9) starting with elements close to the necking zone. Thus, the NR routine aborts at that point as a direct consequence of

Table 3.3.: Results of the elastic strip example.

element type	req. n_{steps}	total n_{NR}	v_{a}
Q1	1	8	2.207
Q1-MIP	1	10	2.207
Q2	1	8	2.194
Q2-MIP	2	16	2.194
Q1/E4	2	12	2.207
Q1/E4-MIP	1	10	2.207
Q1/E4T	1	8	2.208
Q1/E4T-MIP	1	13	2.208
Q1/S5		failure*	

*$u_{\text{p,max}} \approx 5.36$

failure of the material routine. This behavior is almost independent of the number of load steps and mesh refinement. Furthermore, it is not an artifact of intermediate states during the NR procedure, since using the very high number of $n_{\text{steps}} = 6400$ load steps, which leads to almost immediate convergence $n_{\text{NR}} \leq 3$ in every step, fails as well. Even the numerical procedure to invert the stress-strain law proposed by Viebahn et al. [145] finds no solution in those states. All in all, it can be concluded this is an actual mathematical problem of the inverse stress-strain relation of the Neo-Hookean material and not related to the numerical procedures.

All results presented for the 2D case can also be qualitatively observed in 3D simulations.

3.6.4. Thin circular ring

The first 3D example in this work is the thin circular ring shown in Figure 3.11, which was introduced for the analysis of shells by [13] (see also [75, 88]). In the present work, we consider the example for solid elements as described by Korelc et al. [67]. The ring shown in Figure 3.11 has a thickness of $t = 0.03$, an inner radius $r_{\text{i}} = 6$ and an outer radius $r_{\text{o}} = 10$. It is meshed with $2 \times 6 \times 30$ elements. On the fixed face F_1 boundary conditions

$$u(0,0,0) = 0, \quad v(X,Z,Y=0) = 0 \quad \text{and} \quad w(X,Y=0,Z=0) = 0 \tag{3.42}$$

apply. To complete the setup, a surface dead load $q = 6.67 \cdot 10^{-3}$ is applied in z-direction on face F_2 and the elasticity constants are chosen to $\mu = 10500$ and $\lambda = 0$.

Results of this numerical example are summarized in Table 3.4 for a selected set of elements. Note that H1 and H1-MIP are excluded from Table 3.4 due to severe locking, which yields underestimated displacement and thus also few necessary NR iterations (see Section 3.6.2). For all other elements, the MIP method greatly improves convergence of the NR procedure. It enables to apply the complete load for all EAS elements within one step instead of 3-5

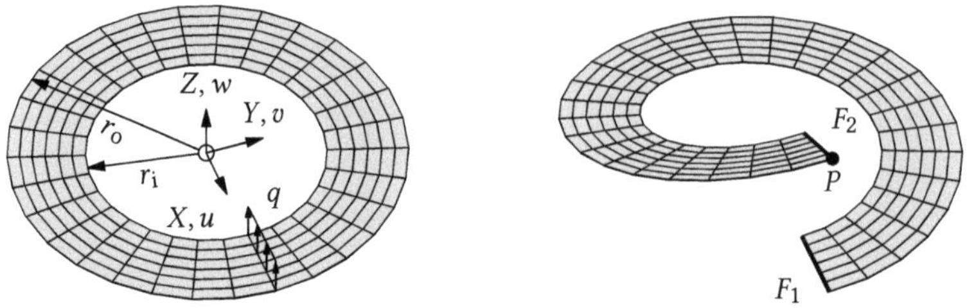

Figure 3.11.: Thin circular ring. Problem setup with undeformed mesh (left) and deformed mesh for $q = 6.67$ computed with H1/E9 (right).

without MIP. Furthermore, the number of total necessary NR-iterations n_{NR} is reduced by a factor of at least 5.9 for the St. Venant-Kirchoff material and 4.7 for the Neo-Hookean model, with the greatest improvement observed for element HA1/E12T. As demonstrated before, AS element H1/S18 shows again superior robustness for both material models and is on top of that closest to the converged result of $w_{\mathrm{P}} = 10.26$.

Table 3.4.: Results of the thin circular ring example.

	St. Venant-Kirchhoff			Neo-Hooke		
element type	req. n_{steps}	total n_{NR}	w_{P}	req. n_{steps}	total n_{NR}	w_{P}
H1/E9	3	41	7.311	4	56	7.311
H1/E9-MIP	1	7	7.311	1	9	7.311
H1/E9T	3	41	7.314	3	42	7.314
H1/E9T-MIP	1	7	7.314	1	9	7.314
HA1/E12T	3	48	8.198	5	81	8.198
HA1/E12T-MIP	1	8	8.198	1	11	8.198
H1/S18	1	7	9.741	1	7	9.741

3.6.5. Spherical shell with opening

The second 3D example is the spherical shell problem with opening shown in Figure 3.12 (see also [6, 29, 67, 95]). The spherical structure with middle radius $r_{\mathrm{m}} = 10$, a thickness of $t = \{0.5, 0.05\}$ and opening angle $\beta = 18°$ is supported by boundary conditions $u(X, Y=0, Z) = 0$, $v(X=0, Y, Z) = 0$ and $w(X, Y, Z=0) = 0$. Prescribed displacements $w(r=r_{\mathrm{i}}, \varphi, \theta=\beta) = 10$ are applied on the lower edge of the opening. The shell is meshed with $2 \times 16 \times 16$ elements and the elasticity constants are chosen to $\lambda = 1.2115 \cdot 10^5$ and $\mu = 8.0769 \cdot 10^4$ (corresponding to $E = 2.1 \cdot 10^5$ and $\nu = 0.3$).

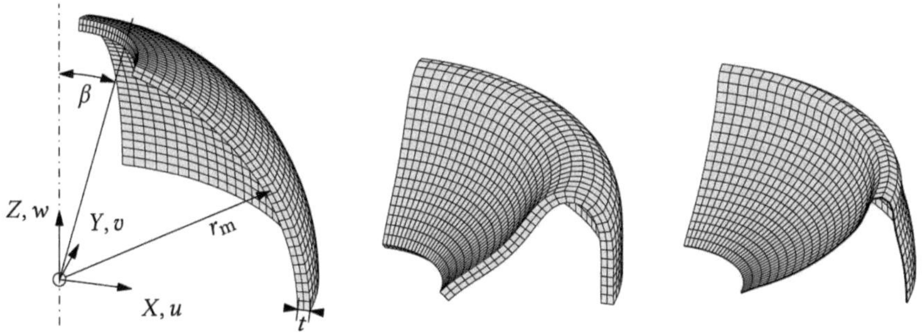

Figure 3.12.: Spherical shell with opening. Problem setup with undeformed mesh (left) and deformed mesh for $w = 10$ computed with H1/E9 with $t = 0.5$ (middle) and $t = 0.05$ (right), respectively.

The required numbers of load steps n_{steps} and NR iteration n_{NR} as well as the reaction force in z-direction R_z are shown in Table 3.5 for both the St. Venant-Kirchhoff and Neo-Hookean material model.

In a first step we examine the results for $t = 0.5$. For this setup, all elements exhibit reaction forces within a 2% margin of the converged result for both materials. More interesting are the required number of load steps and NR iterations. It can be observed for the St. Venant-Kirchhoff material, that, even though the number of load steps differs, almost the same amount of total NR iterations is needed regardless of which element is used. However, for the Neo-Hookean material larger differences can be observed. For this material model the AS element H1/S18 excels by still requiring only a total of $n_{\text{NR}} = 27$ NR iterations. All other elements need at least twice the amount of iterations. Better results are obtained with the MIP modified versions, which require approximately 15% less NR steps compared to the standard EAS elements.

For the thin thickness $t = 0.05$ more load steps and iterations are necessary for all elements and the reaction forces differ substantially. However, improvements due to the MIP method are more pronounced, which is in line with results in Section 3.6.3. Since smaller occurs for the thinner shell, (3.33) is less severely violated which leads to the improved behavior of the MIP method.

All in all, this example confirms that the MIP does not always yield such high improvements as shown in Section 3.6.2 and 3.6.4, depending on the magnitude of the strain. However, it usually improves robustness of strain driven elements if strains are not too high.

Table 3.5.: Results of the spherical shell test.

		St. Venant-Kirchhoff			Neo-Hooke		
t	element type	req. n_{steps}	total n_{NR}	R_z	req. n_{steps}	total n_{NR}	R_z
0.5	H1/E9	2	25	5043	7	69	5371
	H1/E9-MIP	3	25	5043	8	58	5371
	H1/E9T	2	25	5161	9	77	5488
	H1/E9T-MIP	3	25	5161	8	58	5488
	HA1/E12T	2	25	5162	9	77	5490
	HA1/E12T-MIP	3	25	5162	8	58	5490
	H1/S18	2	27	5131	3	27	5454
0.05	H1/E9	8	140	19.28	30	313	19.25
	H1/E9-MIP	8	69	19.28	32	224	19.25
	H1/E9T	8	143	27.99	32	306	27.14
	H1/E9T-MIP	11	100	27.99	29	206	27.14
	HA1/E12T	12	166	28.01	32	306	27.17
	HA1/E12T-MIP	8	77	28.01	29	206	27.17
	H1/S18	7	67	21.58	6	59	21.38

3.6.6. Elasto-plastic circular bar

The final example of the present work is an elasto-plastic simulation based on the material model proposed by Simo [130], which is described in more detail in Appendix 3.A.3. For this model the material parameters are set to the standard values $\mu = 80.1938$, $\kappa = 164.206$, $\sigma_{Y0} = 0.45$, $\sigma_{Y\infty} = 0.715$, $\delta = 16.93$ and $H = 0.12924$ (see, e.g., [8, 47, 67, 105, 129–132, 134]). The test covers necking of a circular bar with radius $R = 6.413$ and a total length of $2L = 53.334$ (see, e.g., [6, 46, 67, 105, 115, 123, 130–133]). Due to symmetry, only one eighth of the bar has to be considered, which is shown in Figure 3.13. To initiate necking, the radius is linearly reduced from R to $\bar{R} = R - 0.07$ along the length of the bar. Two quarter cylinders with 480 elements each are used to mesh the specimen such that the lower fifth of the bar is refined (see Figrue 3.13). Symmetry boundary conditions $u_i(X_i = 0, X_j, X_k) = 0,\ i, j, k \in \{1, 2, 3\}$ apply and the structure is loaded by prescribed displacement $u(X = L, Y, Z) = \bar{u}$. During the simulations $\bar{u}$ is gradually increased to $\bar{u} = 7.0$, where the first half of load steps is used until $\bar{u} = 5.6$ and the second half covers the remaining $\Delta\bar{u} = 1.4$. This is a standard procedure (see, e.g., [105, 131]) since the final steps of this simulation are especially demanding due to the softening of the material.

Usually, a *line-search* (LS) algorithm is used to stabilize the NR procedure in elasto-plastic simulations (see, e.g., [6, 105, 130, 131, 133]). In the present work we use the method described in Bonet and Wood [25].

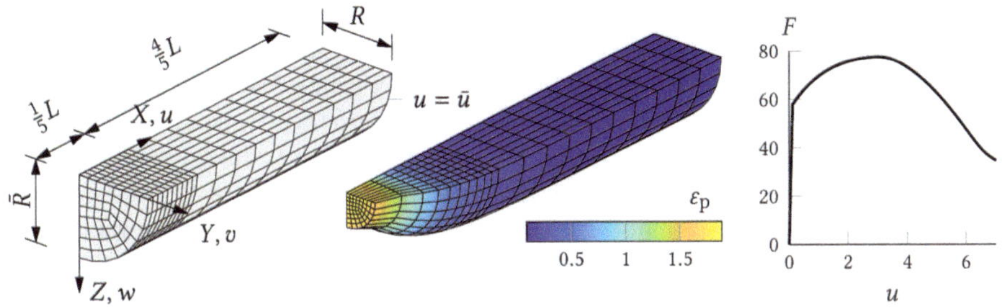

Figure 3.13.: Elasto-plastic circular bar. Reference configuration (left), deformed configuration with distribution of accumulated plastic strain ε_p for $\bar{u} = 7.0$ computed with HA1/E12T using 20 load steps (middle) and load-displacement curve computed with 100 load steps and HA1/E12T (right).

The total number of NR iterations is shown in Table 3.6 for element H1/E9 and various numbers of load steps. Without MIP method and LS the simulation requires at least 28 load steps and is volatile in a sense that it is not guaranteed that a higher number of load steps leads to a solution. The robustness can be improved with either the MIP method or a line search algorithm. Both reduce the number of necessary load steps and ensure (with one exemption) that the NR method converges for higher number of load steps. The best results are obtained by combination of both methods. This allows the lowest number of load steps and also needs slightly fewer iterations than the line search approach without MIP method.

Table 3.6.: Elasto-plastic circular bar test. Total number of NR iterations for H1/E9.

load steps n_{step}	16	18	20	22	24	26	28	30	32	34	36	38	40
H1/E9	-	-	-	-	-	-	176	-	-	209	-	226	233
H1/E9+LS	-	-	148	159	168	178	193	196	220	229	235	243	258
H1/E9-MIP	-	-	-	-	-	169	180	182	201	201	-	216	228
H1/E9-MIP+LS	134	-	146	157	166	180	186	196	220	222	231	242	257

3.7. Conclusion

The present work covers extensions to the MIP method, which has recently been proposed by Magisano et al. [88]. This method has so far been used to improve numerical robustness (meaning number of required NR iterations and size of applicable load steps) of displacement-based and mainly structural finite elements using the St. Venant-Kirchhoff material. Herein, we considered solid finite elements and proposed three extensions.

First of all, we suggested a simple extension to general material models. As for the standard MIP method, we do not interfere with the residuum of the nonlinear FE problem

and only modify the tangent, which means that the converged solution of standard FE simulations and the MIP modified versions are identical. Only the robustness of the MIP modified elements is increased as we showed in many numerical simulations. Second, a MIP version using a spatial formulation of FE equations was introduced in order to simplify implementation of more complex material models such as the elasto-plastic model considered in this work. Finally, we extended the method to EAS finite elements and were able to overcome one of the major drawbacks of these elements in many applications. In particular, we showed that the proposed MIP approach cures their lack of robustness when compared to assumed stress elements. With the simple modifications necessary for the MIP method, EAS elements become much more efficient and robust, which makes them even more interesting for practical simulations.
On top of that, we proposed a novel inverse stress-strain relation for a Neo-Hookean material, which had to the best knowledge of the authors not been proposed before. With this relation we were able to show that AS elements exhibit superior robustness compared to classical strain driven element formulations. Unfortunately, in accordance with literature, we also showed that this approach is not generally applicable in all states of strain. In fact, this observation provides another argument in favor of using much simpler strain-driven finite elements together with the MIP modification. This approach allows the benefit of both, favorable robustness and simple implementation of complex material models.

However, there are still a few open issues. While the MIP method greatly increases robustness of finite elements for general material models and small strain, less improvement or in some special cases even slightly disadvantageous behavior can be observed if large strain occurs. A thorough explanation and cure of this issue should be the goal of further investigations. Another line of research could follow creating a MIP approach based on a *Hu-Washizu* functional instead of the *Hellinger-Reissner* approach followed thus far.

Appendix to Chapter 3

3.A. Material models

This appendix covers all material models employed in the numerical examples in Section 3.5.4 and 3.6 of the present work. To that end two hyperelastic materials and a elasto-plastic material model are summarized subsequently. Special emphasize is put on inverse stress-strain relations needed for AS elements. The only constitutive law that is straightforward to invert is a *St. Venant-Kirchhoff* material model (see Appendix 3.A.1) due to its linear relation between S and E. To the best knowledge of the authors, an analytical inversion of the stress-strain relations exists, apart from the St. Venant-Kirchhoff model, only for a *Neo-Hookean* model, which is described by Wriggers [154] (Chapter 10.3). A different Neo-Hookean model is examined in this work and its inverse stress-strain relation is derived in Section 3.A.2.

We omit accent "$(\hat{\bullet})$", which is used to denote constitutive quantities in the rest of this work, in this appendix in order to simplify notation.

3.A.1. St. Venant-Kirchhoff

The first material model considered is the well-known *St. Venant-Kirchhoff* model. Its strain-energy function is given by

$$W = \frac{1}{2}\mathbf{E} : \mathbb{C}_{\mathrm{SVK}} : \mathbf{E}, \tag{3.43}$$

where $\hat{\mathbb{C}}_{\mathrm{SVK}}$ denotes the constant fourth-order linear elasticity tensor. This quadratic form leads, according to (3.5), to a linear relation between Green-Lagrange strain E and the second Piola-Kirchhoff stress S. It is in 3D[7] vector-matrix form given by

$$\begin{bmatrix} S_{11} \\ S_{22} \\ S_{33} \\ S_{12} \\ S_{23} \\ S_{13} \end{bmatrix} = \begin{bmatrix} 2\mu+\lambda & \lambda & \lambda & 0 & 0 & 0 \\ \lambda & 2\mu+\lambda & \lambda & 0 & 0 & 0 \\ \lambda & \lambda & 2\mu+\lambda & 0 & 0 & 0 \\ 0 & 0 & 0 & \mu & 0 & 0 \\ 0 & 0 & 0 & 0 & \mu & 0 \\ 0 & 0 & 0 & 0 & 0 & \mu \end{bmatrix} \begin{bmatrix} E_{11} \\ E_{22} \\ E_{33} \\ 2\,E_{12} \\ 2\,E_{23} \\ 2\,E_{13} \end{bmatrix}, \tag{3.44}$$

where μ and λ are the *Lamé* constants. Note furthermore that (3.44) is straightforward to invert by simply computing the inverse of the constant $\mathbb{C}_{\mathrm{SVK}}$ in vector-matrix form. Thus, the St. Venant-Kirchhoff material can easily be used for AS-elements introduced in Section 3.2.2.3.

3.A.2. Neo-Hooke

3.A.2.1. Standard form

The second material law considered in this work is a *Neo-Hooke* law with strain-energy function

$$W = \frac{\mu}{2}\left(\mathrm{tr}(\mathbf{C}) - 3\right) + \frac{\lambda}{2}\ln^2 J - \mu \ln J, \tag{3.45}$$

[7] The 2D plane strain case is obtained by simply crossing out the corresponding rows and columns.

where $\mu > 0$ and $\lambda > -2/3\mu$ are the *Lamé* constants and $J = \det \mathbf{F}$. Derivation of this definition with respect to $\mathbf{C}$ yields the second Piola-Kirchhoff stress

$$\mathbf{S} = 2\frac{\partial W}{\partial \mathbf{C}} = \mu\left(\mathbf{I} - \mathbf{C}^{-1}\right) + \lambda \ln J \mathbf{C}^{-1}. \tag{3.46}$$

3.A.2.2. Inverse Neo-Hookean law

The inverse relation of (3.46) presented here has to the best knowledge of the authors never been proposed before. Only a similar inverse relation for a different Neo-Hookean model is given in Wriggers [154]. Similar to the derivations there, several cases have to be considered which are laid out subsequently.

Case $\lambda = 0$
In this case, which corresponds to $\nu = 0$, it is straightforward to obtain

$$\mathbf{C} = \left(\mathbf{I} - \frac{1}{\mu}\mathbf{S}\right)^{-1} \tag{3.47}$$

as inverse stress-strain relation from (3.46). Thus, $\mathbf{I} - \mu^{-1}\mathbf{S}$ must be invertible which can be examined with relations $\det(\mathbf{I}+\mathbf{A}) = 1 + I_1(\mathbf{A}) + I_2(\mathbf{A})$ and $\det(\mathbf{I}+\mathbf{A}) = 1 + I_1(\mathbf{A}) + I_2(\mathbf{A}) + I_3(\mathbf{A})$ holding for an arbitrary tensor $\mathbf{A}$ with Invariants $I_i(\mathbf{A})$ in 2D and 3D, respectively (see, e.g., [103]). From these relations, condition $\det(\mathbf{I} - \mu^{-1}\mathbf{S}) \neq 0$ for invertibility and requirement $J^2 = \det(\mathbf{C}) > 0$ necessary for physically meaningful results, we get that the eigenvalues λ_i^S of $\mathbf{S}$ must fulfill either $\lambda_1^S, \lambda_2^S > \mu$ or $\lambda_1^S, \lambda_2^S < \mu$ in the 2D case. For 3D problems either restrictions $\lambda_i^S, \lambda_j^S > \mu$, $\lambda_k^S < \mu$ or $\lambda_i^S, \lambda_j^S, \lambda_k^S < \mu$ where i, j, k are permutations of $\{1, 2, 3\}$ apply. See also Ogden [98] Chapter 6.2.2 for similar results.

Case $\lambda \neq 0$
In this case simple rearranging of (3.46) yields

$$\frac{1}{\lambda}\left(\mu\mathbf{I} - \mathbf{S}\right) = \left(\frac{\mu}{\lambda} - \ln J\right)\mathbf{C}^{-1} \quad \overset{a\neq 0}{\Leftrightarrow} \quad \mathbf{C} = \beta\mathbf{A}^{-1} \tag{3.48}$$

where auxiliary variables

$$\mathbf{A} = \frac{1}{\lambda}\left(\mu\mathbf{I} - \mathbf{S}\right), \quad a = \det(\mathbf{A}), \quad \beta = \left(\frac{\mu}{\lambda} - \ln J\right) \tag{3.49}$$

have been introduced. Tensor $\mathbf{A}$ has to be invertible which implies that the eigenvalues of $\mathbf{S}$ have to fulfill the same requirements as in the case $\lambda = 0$. The inverse stress-strain relation of (3.46) is given by $(3.48)_2$ where the only unknown is $\beta(J)$.

In case of $a = 0$ $(3.48)_1$ yields $\beta = 0$ since $\mathbf{C}$ has to be invertible for physically meaningful results. Thus only $\mathbf{A} = \mathbf{0} \Leftrightarrow \mathbf{S} = \mu\mathbf{I}$ would be allowed in order to fulfill $(3.48)_1$. In that case, however, C is not uniquely defined and thus no inversion of the stress-strain relation is possible. If $a \neq 0$, the next step is to take the determinant of $(3.48)_1$. This yields for both the 2D plane strain and 3D case an equation for J given by

$$\beta^d - aJ^2 = 0, \tag{3.50}$$

where $d \in \{2, 3\}$ is the spatial dimension. Lengthy computations using the *Lambert-W* function $\mathcal{W}(x)$ finally yield

$$\ln J = \frac{\lambda}{\mu} - \frac{d}{2}\mathcal{W}\left(\frac{a}{|a|}\exp\left(\frac{2}{d}\frac{\mu}{\lambda} - \ln\frac{d}{2} + \frac{1}{d}\ln|a|\right)\right), \tag{3.51}$$

where requirements

$$a \neq 0, \qquad a > \begin{cases} 0, & d = 2 \\ -(\frac{d}{2})^d \exp\left(-d - 2\frac{\mu}{\lambda}\right), & d = 3 \end{cases} \tag{3.52}$$

have to be met. Note that always branch $\mathcal{W}_0$ of the Lambert-W function is needed, which follows from the solution for $a > 0$ and continuity requirements. Furthermore, (3.50) automatically ensures $J > 0$ which allows using the logarithm in (3.51).

Remark 3.9. *Note that regardless of the case $\lambda = 0$ or $\lambda \neq 0$ there are restrictions on which stress tensors* S *allow a unique inversion of the stress-strain relation. These states actually occur in practical simulations with large strain as shown in Section 3.6.3. In that case these states lead to failure of the computation. This is* not *a problem of the numerical procedure but results form the physical equations i.e.,* (3.46) *as shown in Section 3.6.3.*

3.A.3. Logarithmic strain based von Mises elasto-plasticity

The final material model considered in the present work is the *elasto-plastic* model proposed by Simo [130]. This eigenvalue based formulation is widely used in the context of finite element development (see, e.g., [8, 65, 105, 132]) and based on the standard multiplicative split $\mathbf{F} = \mathbf{F}^e\mathbf{F}^p$ into elastic and plastic parts. Its elastic response is governed by a *Hencky* strain-energy function which employs the logarithmic principal stretches. The plastic part of the model is governed by the *von Mises yield condition* with *nonlinear isotropic hardening with saturation* and the *associative flow rule*. More information on the material model and algorithms for standard elements are given in the work of Simo [130]. For the numerical implementation of the model an eigenvalue perturbation technique according to Miehe [92] is applied to avoid treatment of duplicate stretches.

3.B. Two DOF example

This appendix gives additional details on the simple 2 DOF example presented in Section 3.3. Focus is put on the assumptions and derivations needed to obtain (3.19) and (3.20). First of all, the kinematic relation in (3.18) follows from the assumption of linear displacements u and v along the axis of the bar. This implies that the Green-Lagrange strain is constant and given by

$$\varepsilon = \frac{1}{2}\frac{(l^2 - L^2)}{L^2} = \frac{1}{2}\frac{\left(\sqrt{(L+u)^2 + w^2}\right)^2 - L^2}{L^2} \overset{L=1}{=} u + \frac{1}{2}(u^2 + w^2), \tag{3.53}$$

where $L = 1$ and l are the original and deformed length of the bar, respectively. From the assumption of a linear relation between Green-Lagrange strain and the normal force N (i.e., St. Venant-Kirchhoff material) we get the constitutive relation

$$\hat{N} = EA\varepsilon, \tag{3.54}$$

where A is the constant reference cross section of the bar and E denotes its Young's modulus. The variational functional associated with the 2 DOF example can be cast in the form

$$\Pi^{2\text{DOF}} = \Pi^{\text{bar}} + \frac{1}{2}kw^2 + \lambda u - c\lambda w. \tag{3.55}$$

Therein, k is the stiffness of the linear spring and λ is the external force. The internal potential of the bar Π^{bar} is given by

$$\Pi_{\text{U}}^{\text{bar}} = \frac{1}{2}\hat{N}\varepsilon L \qquad \text{and} \qquad \Pi_{\text{AS}}^{\text{bar}} = N\left(\varepsilon - \frac{1}{2}\hat{\varepsilon}\right)L \tag{3.56}$$

in the displacement-based and AS case, respectively. For the AS formulation an independent stress field N is introduced. Furthermore, the *Legendre*-transformation of the internal energy $W(\varepsilon) = \hat{N}\varepsilon$ given by $U(\sigma) = N\hat{\varepsilon} - W(\hat{\varepsilon})$ has been used. Therein, $\hat{\varepsilon}$ is strain obtained from the inverse stress strain relation. Variation of the internal part Π^{bar} of functional $\Pi^{2\text{DOF}}$ yields

$$\delta\Pi_{\text{U}}^{\text{bar}} = L\hat{N}(\varepsilon)\delta_{\boldsymbol{u}}\varepsilon \qquad \text{and} \qquad \delta\Pi_{\text{AS}}^{\text{bar}} = LN\delta_{\boldsymbol{u}}\varepsilon + L\delta N(\varepsilon - \hat{\varepsilon}) \tag{3.57}$$

for the displacement-based and AS formulation. Variations (3.57) are closely related to the continuum formulations (3.7) and (3.15). In fact, in case of the bar, we have a uniaxial stress state, where the only non-zero stress component is S_{11}, and furthermore constant stress and strain along the bar. Imposing these restrictions on (3.7) and (3.15) directly yields (3.57). Note that this close relation allows the transfer of the results concerning robustness from the simple bar problem to the continuum formulation.

Imposing the stationary condition $\delta\Pi^{2\text{DOF}} = 0$ and subsequent linearization yields the residuum and tangent of both the displacement-based and AS form of the 2 DOF example given in (3.19) and (3.20) respectively.

4. Mesh distortion insensitive and locking-free Petrov-Galerkin EAS elements for linear elasticity

This chapter reproduces:*
Pfefferkorn R and Betsch P. "Mesh Distortion Insensitive and Locking-Free Petrov-Galerkin Low-Order EAS Elements for Linear Elasticity". In: *Int J Numer Meth Eng*. 122(23): 6924–6954, 2021. DOI: 10.1002/nme.6817

Abstract: One of the most successful mixed finite element methods in solid mechanics is the enhanced assumed strain (EAS) method developed by Simo and Rifai in 1990 [134]. However, one major drawback of EAS elements is the highly mesh dependent accuracy. In fact, it can be shown that not only EAS elements, but every finite element with a symmetric stiffness matrix must either fail the patch test or be sensitive to mesh distortion in bending problems (higher order displacement modes) if the shape of the element is arbitrary. This theorem was established by MacNeal in 1992 [84].
In the present work we propose a novel Petrov-Galerkin approach for the EAS method, which is equivalent to the standard EAS method in case of regular meshes. However, in case of distorted meshes, it allows to overcome the mesh distortion sensitivity without loosing other advantages of the EAS method. Three design conditions established in this work facilitate the construction of the element which does not only fulfill the patch test but is also exact in many bending problems regardless of mesh distortion and has an exceptionally high coarse mesh accuracy. Consequently, high quality demands on mesh topology might be relaxed.

Keywords: enhanced assumed strain (EAS), mesh distortion, Petrov–Galerkin, linear elasticity, skew coordinates

4.1. Introduction

In the early days of the finite element method it was soon discovered that *low-order displacement-based* elements severely underestimate displacement under many circumstances such as in bending dominated problems and the incompressible limit. This phenomenon was termed *locking* and prohibits reasonable utilization of low-order displacement-based elements in engineering applications (e.g., MacNeal [85]). Thus, a plethora of remedies has been developed since the 1960s which can essentially be grouped into three main categories: higher order methods [34, 85, 162], reduced integration with stabilization [14, 16, 40] and mixed finite elements [110, 134, 137, 141]. All of these remedies lead to substantially improved finite elements and some of each category are available in commercial software.

Despite the tremendous effort put into developing new finite elements and enhancing their performance, there have been hardly any major breakthroughs in classical methods since the mid 1990s (with the exemption of the isogeometric analysis, see [34]). An explanation for this can be found in a landmark paper published by MacNeal [84] in 1992 (see also the preliminary work [83]), which has in the opinion of the authors not gotten the attention it deserves. In this work, MacNeal proves the theorem that a finite element with arbitrary shape *cannot* simultaneously satisfy the constant strain patch test [162] and be exact for higher order modes under the premise of a symmetric stiffness matrix. This fundamental limit to an element's perfectibility applies regardless of internal methodologies, that is, which of the remedies to cure locking is used. Thus, the final conclusion drawn by MacNeal [84] is that it is impossible to substantially improve elements beyond what has already been achieved with the standard (Bubnov-Galerkin) approach. Indeed, to the best knowledge of the authors, no element could so far break the limits of MacNeal's theorem.[1]
In particular, most elements have symmetric stiffness matrices, which come of course with many advantages.[2] Moreover, they are usually designed to fulfill the patch test, see, e.g., [110, 134], since this comes with a lot of benefits (see [162]) as well. Consequently, in accordance with MacNeal's theorem, they perform poorly in (higher order) bending problems if meshes are distorted. Vice versa, the less frequent choice of (deliberately or not) sacrificing the ability to pass the patch test enables construction of elements with increased bending accuracy. An example for this approach is the incompatible mode model by Wilson et al. [149], which famously fails the patch test, and the recently proposed method of reverse adjustment to the patch test by Hu et al. [54, 55] (which ironically does not pass the strong patch test).
All in all, improvement of element performance beyond what has already been achieved almost inevitably leads to an unsymmetric stiffness matrix. MacNeal even briefly mentions this possibility but then deems this option "abhorrent for many reasons" [84]. However, the works of, e.g., Rajendran et al. [116, 117], Xie et al. [158] and the present contribution show that there is much to be gained with unsymmetric stiffness matrices.

[1] Cen et al. [32] claim to break through MacNeal's theorem. However, their element has an unsymmetric stiffness matrix and does therefore not violate the limits proven by MacNeal [84]. Another interesting candidate is the element by Wu and Chueng [157]. It has a symmetric stiffness matrix and is exact in a bending problems at some nodes. Unfortunately, the displacement is not exact at all nodes and it does thus also not break through MacNeal's theorem (see Sze [139]).

[2] E.g., the reduced computational cost to solve the linear equation system and decreased memory consumption since only half of the sparse matrix has to be stored.

The first element with unsymmetric stiffness matrix has been proposed by Rajendran and Liew [117], who chose a *Petrov-Galerkin* approach for higher order displacement elements instead of the usual *Bubnov-Galerkin* ansatz, and named their method unsymmetric finite element method. The key idea is to use *metric* [85, 118] shape functions, which are constructed in the physical space, as ansatz for the trial function of the displacement while the usual isoparametric functions are employed for the test function. Unfortunately, merely using physical coordinates leads to frame-dependent ansatz functions as noted by Ooi et al. [100]. A cure for this issue, which does not induce other problems such as anisotropies, has been proposed by Xie et al. [158]. It involves skew coordinates, which are affine-equivalent [119] to physical coordinates and objective. The skew frame has first been proposed for assumed stress finite elements (see Yuan et al. [159] and Wisniewski and Turska [150]) and has subsequently also been employed for other mixed elements [113, 114, 150–152] and unsymmetric finite elements [59, 158, 161].
While higher order unsymmetric finite elements work well and pass higher order patch tests regardless of element shape (see [56, 99, 101, 116, 117, 158]), there is a fundamental issue for low-order elements: Consider, e.g., a plane four-node quadrilateral element with eight displacement degrees of freedom. It is then impossible to represent complete quadratic polynomials since these would require at least twelve degrees of freedom. Thus, it is crucial to choose the higher order modes carefully in order to get the best performance. Consequently, the unsymmetric low-order elements presented so far have either extremely complex ansatz functions which are material dependent [32, 76, 78, 158, 161] or require higher order integration and many internal degrees of freedom [59]. Furthermore, the elements are often not straightforward to extend to the 3D case.

However, the advantages of low-order elements with regard to mesh generation, (stress) singularities, and bandwidth (sparsity) of the stiffness matrix make low-order unsymmetric finite elements desirable. Hence, we propose a novel low-order unsymmetric mixed element based on a *Petrov-Galerkin* approach for the *enhanced assumed strain* (EAS) method introduced by Simo and Rifai [134]. We will show that this approach allows to construct low-order unsymmetric elements without the drawbacks described above.
We choose to start our developments from the EAS method since the commonly used symmetric (Bubnov-Galerkin) version of it features several desirable properties: It fulfills the patch test, is locking-free in case of undistorted meshes, and, most importantly, it is straightforward to extended to large deformations and general material models due to its strain-driven format (see Simo et al. [131, 132] and Glaser and Armero [47]). Consequently, it is one of the most frequently used mixed methods in research and application (see, e.g., [3, 8, 47, 66, 67, 80, 82, 104, 105, 109, 113, 114, 131, 132, 134]). With the newly proposed Petrov-Galerkin approach we show that it is possible to overcome the sensitivity to mesh distortion of existing EAS elements.
To that end, we first establish *three design conditions* required in order for the element to be exact for a specific displacement mode. If these conditions are met, the element is exact for that mode in the sense that the nodal displacements coincide with the analytic solution (nodally exact response). In particular, we choose to fit the novel element to a modified version of the assumed stress modes proposed by Pian and Sumihara [110] and

Pian and Tong [112] in 2D and 3D, respectively. These stress modes include besides the important patch test modes also bending modes which are of the utmost importance in many engineering applications. The resulting element is therefore exact for constant stress and bending problems *regardless* of element shape. It is also locking-free, frame-indifferent, isotropic and its stiffness matrix is integrated exactly by the standard Gauss quadrature rule. Moreover, the newly proposed unsymmetric EAS element exhibits an substantially increased coarse mesh accuracy. Finally, in case of regular meshes, the element coincides with the original well working EAS element by Simo and Rifai [134].
The design of the novel element with all these desirable properties is made possible by combining ingredients from a multitude of previously developed element formulations. Besides the obvious *EAS framework* [132, 134] we also employ ideas from *assumed stress* approaches [110, 112, 150], the *skew coordinate* frame [150, 158, 159], *incompatible mode* elements [22, 59, 141, 149] and others [16, 62, 151].
Naturally, the unsymmetric incompatible mode 2D element proposed by Huang et al. [59] is closely related to the present Petrov-Galerkin EAS element. However, the novel EAS approach is more general, requires less internal degrees of freedom and allows to examine the underlying mechanisms in a deeper way. Interestingly, a violation of one of the three design conditions for exact solutions mentioned above is the reason why the 2D element by Huang et al. [59] is not straightforward to extend to 3D.

The present work is structured into six sections. In Section 4.2 we revisit MacNeal's theorem since it is key to the methods developed in the remainder of this work. We aim at giving a simpler approach to the proof in [84] and present some extensions to the original theorem of MacNeal. To that end we compare the finite element approximation to the continuum description and examine both formulations in detail in Sections 4.2.1 and 4.2.2, respectively. After that, conclusions of MacNeal's theorem are drawn in Section 4.2.3. Some generalizations of the proof in Section 4.2 can be found in Appendix 4.A and the key findings of MacNeal's theorem are summarized in Section 4.2.4. Section 4.3 covers the weak form for EAS elements and the three design conditions required for nodally exact solutions. Afterwards, we determine ansatz spaces which fulfill the design conditions in Section 4.4. To that end we first introduce the skew coordinate frame in Section 4.4.1 and describe the analytic modes for which the element is optimized in Section 4.4.2. The actual ansatz spaces for the Petrov-Galerkin EAS elements are covered in Section 4.4.3 for the 2D case and in Section 4.4.4 for the 3D case, respectively. Numerical simulations comparing the novel element to established ones are presented in Section 4.5 before conclusions are drawn in Section 4.6.

4.2. MacNeal's theorem

MacNeal's theorem [84] states that a finite element *cannot* simultaneously pass the patch test (see, e.g., [162]) and be exact for higher order displacement modes in case of arbitrarily distorted meshes if its stiffness matrix is *symmetric*.

To prove this proposition we consider, following MacNeal [84], a linear elastic continuum $\mathcal{B}$ and a single finite element $\Omega^e \subset \mathcal{B}$ of arbitrary shape and arbitrary number of nodes N which is embedded into $\mathcal{B}$. The two domains interact at the element's boundary $\partial\Omega^e$. This connection is described by displacement $\mathbf{u}(\mathbf{x})$ and traction $\mathbf{t}(\mathbf{x})$ for $\mathbf{x} \in \partial\Omega^e$, where $\mathbf{x}$ are Cartesian (or physical) coordinates. Comparing the continuum solutions[3] $\mathbf{u}^*$ and $\mathbf{t}^*$ with the finite element aliases $\mathbf{u}^{\mathrm{h},e}$ and $\mathbf{t}^{\mathrm{h},e}$ on $\partial\Omega^e$ allows to evaluate how well the element replaces the respective continuum domain[4]. To that end we examine both formulations in depth in the sequel.

Example Q1. *Figure 4.1 shows exemplarily the case of a plane quadrilateral element with straight edges and four nodes (Q1). This simple case is used throughout the remainder of this work to clarify relations.*

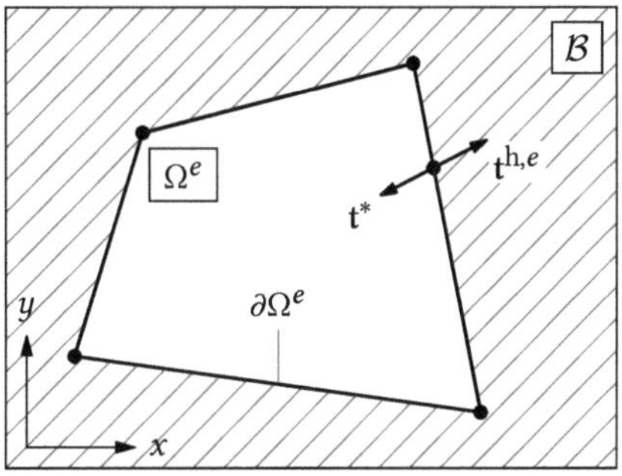

Figure 4.1.: Single finite element Ω^e embedded into linear elastic continuum $\mathcal{B}$.

4.2.1. Continuum domain

4.2.1.1. Linear elasticity

Given a prescribed displacement field $\mathbf{u}^*(\mathbf{x})$ for $\mathbf{x} \in \mathcal{B}$, the usual relations for a linear elastic continuum

$$\boldsymbol{\varepsilon}^* := \nabla^s_{\mathbf{x}} \mathbf{u}^*, \tag{4.1a}$$

$$\boldsymbol{\sigma}^* := \mathbb{C} : \boldsymbol{\varepsilon}^*, \tag{4.1b}$$

$$\mathbf{t}^* := \boldsymbol{\sigma}^* \mathbf{n}, \tag{4.1c}$$

determine the corresponding strain $\boldsymbol{\varepsilon}^*$, stress $\boldsymbol{\sigma}^*$ and traction $\mathbf{t}^*$. Therein, $\mathbb{C}$ denotes the symmetric positive definite fourth order elasticity tensor, $\nabla^s_{\mathbf{x}}(\bullet)$ symbolizes the symmetric

[3] Here and subsequently we frequently omit arguments of functions in order to improve readability.
[4] Throughout the remainder of the work we denote continuum solutions with an asterisk $(\bullet)^*$ and finite element approximations with superscript $(\bullet)^{\mathrm{h}}$.

part of the gradient and $\mathbf{n}$ is the unit outward normal on a surface. Furthermore, the strong form of equilibrium is given by

$$\operatorname{div} \boldsymbol{\sigma}^* + \mathbf{b}^* = \mathbf{0}, \tag{4.2}$$

where $\mathbf{b}^*$ is the field of body force which is readily determined by the prescribed displacement field $\mathbf{u}^*$ via (4.1) and (4.2). By virtue of the given displacement $\mathbf{u}^*$ and (4.1), it is also straightforward to compute appropriate Dirichlet boundary conditions $\bar{\mathbf{u}}^* = \mathbf{u}^*$ on $\partial\mathcal{B}_\mathrm{u} \subset \partial\mathcal{B}$ along with appropriate Neumann boundary conditions $\bar{\mathbf{t}}^* = \mathbf{t}^*$ on $\partial\mathcal{B}_\mathrm{t} \subset \partial\mathcal{B}$ on the body's boundary $\partial\mathcal{B}$. Here, the standard conditions $\partial\mathcal{B}_\mathrm{u} \cap \partial\mathcal{B}_\mathrm{t} = \emptyset$ and $\partial\mathcal{B}_\mathrm{u} \cup \partial\mathcal{B}_\mathrm{t} = \partial\mathcal{B}$ apply. The corresponding *weak form* (or *principle of virtual work*) is given by

$$\int_{\mathcal{B}} \nabla^s_\mathbf{x} \mathbf{v}^* : \boldsymbol{\sigma}^* \, \mathrm{d}V = \int_{\mathcal{B}} \mathbf{v}^* \cdot \mathbf{b}^* \, \mathrm{d}V + \int_{\partial\mathcal{B}_\mathrm{t}} \mathbf{v}^* \cdot \bar{\mathbf{t}}^* \, \mathrm{d}A \qquad \forall \mathbf{v}^* \in \mathcal{V}, \tag{4.3}$$

where $\mathbf{u}^* \in \mathcal{U} = \{(\mathbf{u})_i \in H_1 |\, \mathbf{u} = \bar{\mathbf{u}}^* \text{ on } \partial\mathcal{B}_\mathrm{u}\}$ is the displacement field introduced above and $\mathbf{v}^* \in \mathcal{V} = \{(\mathbf{v})_i \in H_1 |\, \mathbf{v} = \mathbf{0} \text{ on } \partial\mathcal{B}_\mathrm{u}\}$ is an arbitrary test function. Moreover, in the sequel the left and right hand side of (4.3) are abbreviated by G^*_int and G^*_ext and identify the internal and external part of the weak form, respectively.

4.2.1.2. Displacement modes on Ω^e

Having summarized the basic relations we turn now to the response of the continuum in subdomain Ω^e. Every non-singular displacement state can be represented uniquely by an infinite sum of weighted linearly independent polynomial elementary modes $\mathbf{u}^*_m(\mathbf{x})$. In ascending order these displacement modes are classified as rigid body modes (no strain), patch test modes with constant strain (linear displacement), quadratic displacement modes, and so forth.

Since finite elements have only a specific number of degrees of freedom N_DOF, the discrete solution can only represent a limited amount of modes. Thus, we restrict the following investigation to a linear combination of M linearly independent modes $\mathbf{u}^*_m$, $m = 1, \ldots, M$ with weights α_m such that the displacement and traction are given by

$$\mathbf{u}^* = \sum_{m=1}^{M} \mathbf{u}^*_m \alpha_m \qquad \text{and} \qquad \mathbf{t}^* = \sum_{m=1}^{M} \mathbf{t}^*_m \alpha_m. \tag{4.4}$$

The traction modes $\mathbf{t}^*_m$ relate to $\mathbf{u}^*_m$ via (4.1). Furthermore, to keep expressions as simple as possible, we consider only modes for which

$$\mathbf{b}^* = \mathbf{0} \qquad \Leftrightarrow \qquad \operatorname{div} \boldsymbol{\sigma}^* = \mathbf{0}. \tag{4.5}$$

This includes the important cases of rigid body and constant strain modes, which are characterized by a complete linear displacement field. Furthermore, pure bending modes also fulfill (4.5) [62]. The case $\mathbf{b}^* \neq \mathbf{0}$ is complemented in Appendix 4.A.1.

By considering the usual *Bubnov-Galerkin* approach (see Appendix 4.A.2 for the generalization to a Petrov-Galerkin scheme), which implies that the test function $\mathbf{v}^* = \sum_{m=1}^{M} \mathbf{u}_m^* \beta_m$ use the same modes $\mathbf{u}_m^*$ as the prescribed displacement field $\mathbf{u}^*$, the external part of the weak form on subdomain Ω^e assumes the form

$$G_{\text{ext}}^{*,e} = \int_{\partial\Omega^e} \mathbf{v}^* \cdot \mathbf{t}^* \, \mathrm{d}A = \sum_{m=1}^{M} \beta_m \int_{\partial\Omega^e} \mathbf{u}_m^* \cdot \mathbf{t}^* \, \mathrm{d}A = \sum_{m,n=1}^{M} \beta_m F_{mn}^* \alpha_n = \sum_{m=1}^{M} \beta_m f_m^*. \tag{4.6}$$

Note that we assume that the traction boundary condition applies to the whole boundary $\partial\Omega^e$ of the element, that is, $\partial\Omega_t^e = \partial\Omega^e$, and that the traction is defined by the continuum value $\mathbf{t}^*$, which will be crucial in Section 4.2.2.2. In the equation above, we introduced the generalized modal forces f_m^* and corresponding matrix components F_{mn}^* $(m, n = 1, \dots, M)$ given by

$$f_m^* = \sum_{n=1}^{M} F_{mn}^* \alpha_n \qquad \text{and} \qquad F_{mn}^* = \int_{\partial\Omega^e} \mathbf{u}_m^* \cdot \mathbf{t}_n^* \, \mathrm{d}A. \tag{4.7}$$

Proceeding analogously with the internal part of the weak form yields

$$G_{\text{int}}^{*,e} = \int_{\Omega^e} \nabla_{\mathbf{x}}^s \mathbf{v}^* : \mathbb{C} : \nabla_{\mathbf{x}}^s \mathbf{u}^* \, \mathrm{d}V = \sum_{m,n=1}^{M} \beta_m K_{mn}^* \alpha_n, \tag{4.8}$$

where the components K_{mn}^* constitute the modal stiffness matrix. The major symmetry of $\mathbb{C}$ implies the symmetry of K_{mn}^*. Since the weak form (4.3) holds for every subdomain of $\mathcal{B}$, relation

$$G_{\text{int}}^{*,e} = G_{\text{ext}}^{*,e} \qquad \Leftrightarrow \qquad \int_{\Omega^e} \nabla_{\mathbf{x}}^s \mathbf{v}^* : \mathbb{C} : \nabla_{\mathbf{x}}^s \mathbf{u}^* \, \mathrm{d}V = \int_{\partial\Omega^e} \mathbf{v}^* \cdot \mathbf{t}^* \, \mathrm{d}A \tag{4.9}$$

needs to be satisfied. Now the arbitrariness of β_m yields $K_{mn}^* = F_{mn}^*$. In particular, this establishes the symmetry of F_{mn}^*. Finally, by use of *Gram-Schmidt* orthogonalization

$$\hat{\mathbf{u}}_m^* = \mathbf{u}_m^* - \sum_{k=1}^{m-1} \frac{\int_{\Omega^e} \nabla_{\mathbf{x}}^s \hat{\mathbf{u}}_k^* : \mathbb{C} : \nabla_{\mathbf{x}}^s \mathbf{u}_m^* \, \mathrm{d}V}{\int_{\Omega^e} \nabla_{\mathbf{x}}^s \hat{\mathbf{u}}_k^* : \mathbb{C} : \nabla_{\mathbf{x}}^s \hat{\mathbf{u}}_k^* \, \mathrm{d}V} \hat{\mathbf{u}}_k^*, \tag{4.10}$$

any set of displacement modes $\mathbf{u}_m^*$ can be uncoupled such that matrix F_{mn}^* can be made diagonal. In the following we assume that the modes $\mathbf{u}_m^*$ are uncoupled and thus F_{mn}^* is diagonal. A final conclusion drawn from (4.8) is that $f_m^* = 0$ for rigid body modes $\mathbf{u}_m^{*r}$ since $\boldsymbol{\varepsilon}_m^{*r} = \nabla_{\mathbf{x}}^s \mathbf{u}_m^{*r} = \mathbf{0}$ holds.

4.2.2. Finite element domain

4.2.2.1. Connection between nodes and modes

In the next step we consider the finite element approximation. In contrast to the continuum description, the finite element approximation is based on *nodes* rather than *modes*. It is therefore crucial to determine the relation between the node based finite element displacement $\mathbf{u}^{\mathrm{h},e}$ and the displacement modes introduced in (4.4). For standard (interpolatory) finite elements the displacement on element level is given by

$$\mathbf{u}^{\mathrm{h},e}(\mathbf{x}) = \sum_{i=1}^{N} M_i^e(\mathbf{x})\mathbf{u}_i^e, \tag{4.11}$$

where N is the number of nodes and $\mathbf{u}_i^e$ is the displacement at node i. The corresponding scalar ansatz functions[5] M_i^e, which may (for now) differ from element to element, have the *Kronecker-delta property* $M_i^e(\mathbf{x}_j) = \delta_{ij}$, where $\mathbf{x}_j$ is the position of node j. With these definitions at hand, the relation between modes and nodes can be described by

$$\mathbf{u}_i^e = \sum_{m=1}^{M} \mathbf{u}_{im}^{*,e}\alpha_m, \tag{4.12}$$

where

$$\mathbf{u}_{im}^{*,e} = \mathbf{u}_m^*(\mathbf{x}_i) \tag{4.13}$$

is the value of mode $\mathbf{u}_m^*$ at node $\mathbf{x}_i$. This ensures that the finite element interpolation $\mathbf{u}^{\mathrm{h},e}$ (4.11) coincides with the prescribed displacement field $\mathbf{u}^*$ at the nodes, that is, $\mathbf{u}^{\mathrm{h},e}(\mathbf{x}_j) = \mathbf{u}^*(\mathbf{x}_j)$. However, at other points $\mathbf{x} \in \Omega^e$ the values do generally not coincide.

Remark 4.1. *Assuming a displacement field of the form* (4.11) *is rather restrictive. In general,* M_i^e *are not necessarily scalar and the weights* $\mathbf{u}_i^e$ need not *correspond to the nodal displacements. Such non-interpolatory approaches include the widely used* hierarchical *higher order elements [162] and the* isogeometric *analysis [34]. These generalizations are covered in Appendix 4.A.3.*

4.2.2.2. General finite element framework

Regardless of internal element methodologies, the relation between nodal displacements $\mathbf{u}_i^e$ and nodal forces $\mathbf{P}_i^{\mathrm{h},e}$ of any linear finite element can be written as

$$\sum_{j=1}^{N} \mathbf{K}_{ij}^{\mathrm{h},e}\mathbf{u}_i^e = \mathbf{P}_i^{\mathrm{h},e} \qquad i = 1, \ldots, N, \tag{4.14}$$

[5] In anticipation of the Petrov-Galerkin approach proposed in the sequel we denote the nodal shape functions in (4.11) by M_i^e instead of the more common N_i^e.

where $\mathbf{K}_{ij}^{\mathrm{h},e}$ denotes a partition of the element's *nodal* stiffness matrix $\mathbf{K}^{\mathrm{h},e}$. This format is very general and includes displacement-based elements, reduced integrated elements as well as mixed methods with internal degrees of freedom[6]. The left-hand side of (4.14) refers to the internal part associated with elastic deformations while the right-hand side contains the external nodal forces. Defining the element's *modal* stiffness matrix and *modal* external forces by

$$\tilde{K}_{mn}^{\mathrm{h},e} := \sum_{i,j=1}^{N} (\mathbf{u}_{im}^{*,e})^{\mathrm{T}} \mathbf{K}_{ij}^{\mathrm{h},e} \mathbf{u}_{jn}^{*,e}, \tag{4.15a}$$

$$f_m^{\mathrm{h},e} := \sum_{i=1}^{N} (\mathbf{u}_{im}^{*,e})^{\mathrm{T}} \mathbf{P}_i^{\mathrm{h},e}, \tag{4.15b}$$

allows to recast (4.14) by virtue of (4.12), pre-multiplication with $(\mathbf{u}_{im}^{*,e})^{\mathrm{T}}$ and summation over i in the form

$$\sum_{n=1}^{M} \tilde{K}_{mn}^{\mathrm{h},e} \alpha_n = f_m^{\mathrm{h},e} \qquad m = 1, \dots, M. \tag{4.16}$$

The last equation allows to determine the maximum number of modes M for which an element can exactly reproduce the continuum response in the sense that the nodal displacements are exact. In order for (4.16) to be solvable $\tilde{K}_{mn}^{\mathrm{h},e}$ must have proper rank. To that end we assume that our finite element fulfills the design imperative of stress-free response to rigid body motions. For the corresponding M_r rigid body modes $f_m^* = 0$ as shown in Section 4.2.1.2 while for all other modes $f_m^* \neq 0$.

For the finite element to correctly represent rigid body motions we thus require accordingly $f_m^{\mathrm{h},e} = 0$ and $\mathrm{rank}[\tilde{K}_{mn}^{\mathrm{h},e}] = M - M_r$. Naturally, the *nodal* stiffness matrix has to correctly account for rigid body motion as well. Its rank is then, provided that there are no spurious instable modes, $N_{\mathrm{DOF}} - M_r$ where $N_{\mathrm{DOF}} = N\,d$ is the number of degrees of freedom and d is the spatial dimension. Rewriting the element's modal stiffness matrix (4.15a) in vector-matrix notation yields $\tilde{\mathbf{K}}^{\mathrm{h},e} = (\mathbf{u}^{*,e})^{\mathrm{T}} \mathbf{K}^{\mathrm{h},e} \mathbf{u}^{*,e}$. Thus, the element's modal stiffness matrix has at best the same rank as the nodal stiffness matrix.[7] Therefore, a finite element can in general exactly represent no more than $M = N_{\mathrm{DOF}}$ modes including M_r rigid body modes.

Example Q1. *In case of the quadrilateral element there are $M_r = 3$ rigid body modes, $N = 4$ nodes and $N_{DOF} = 8$ degrees of freedom which determines $M = 8$. Thus, considering the three rigid body modes, the element can only be adapted to a maximum of five displacement modes. Thereof, three are usually chosen to be the constant stress modes (patch test, represented by linear displacement modes). The two remaining higher order modes should therefore be chosen carefully to get the best performance (see also Cen et al. [32]).*

[6] Internal degrees of freedom can be statically condensed on element level to achieve form (4.14).

[7] Theorem $\mathrm{rank}(\mathbf{AB}) \leq \min(\mathrm{rank}\,\mathbf{A}, \mathrm{rank}\,\mathbf{B})$ holds for arbitrary matrices $\mathbf{A}$ and $\mathbf{B}$ [103].

Remark 4.2. *The total of eight modes for Q1 is especially not sufficient to represent fully quadratic displacement modes, which require a total of twelve modes* ($1, x, y, xy, x^2, y^2$ *in both displacement components). It is therefore futile to try and improve quadrilateral elements to exactly represent fully quadratic displacement fields regardless of the applied technique. This observation is the essence of the* principle of limitation *by Fraeijs de Veubeke [42].*

So far, we focused on the internal part of the weak form leading to the stiffness matrix. We now turn to the external part of the weak form which yields the nodal forces $\mathbf{P}_i^{h,e}$ (4.14). In particular, the nodal forces are related to the traction $\mathbf{t}^{h,e}$ acting on the element's boundary $\partial\Omega^e$. We assume that the analytic traction acts on the finite element, that is, $\mathbf{t}^{h,e} = \mathbf{t}^*$. For the element's response to be nodally exact we ultimately have to show that assumption $\mathbf{t}^{h,e} = \mathbf{t}^*$ is consistent with $\mathbf{u}_i^e = \mathbf{u}^*(\mathbf{x}_j)$ (see Section 4.2.3). For the finite element discretization of the external part of the weak form (4.9) we get

$$G_{\text{ext}}^{h,e} = \sum_{i=1}^{N} \mathbf{v}_i^e \cdot \int_{\partial\Omega^e} N_i^e \mathbf{t}^* \, \mathrm{d}A = \sum_{i=1}^{N} \mathbf{v}_i^e \cdot \mathbf{P}_i^{h,e} \quad \Rightarrow \quad \mathbf{P}_i^{h,e} = \int_{\partial\Omega^e} N_i^e \mathbf{t}^* \, \mathrm{d}A, \tag{4.17}$$

where the approximation of the test functions assumes the form (4.11) with nodal shape functions now denoted by N_i^e. Inserting this result into definition (4.15b) of the element's external modal forces $f_m^{h,e}$ yields

$$f_m^{h,e} = \sum_{n=1}^{M} \sum_{i=1}^{N} (\mathbf{u}_{im}^{*,e})^{\mathrm{T}} \int_{\partial\Omega^e} N_i^e \mathbf{t}_n^* \, \mathrm{d}A \alpha_n = \sum_{n=1}^{M} F_{mn}^{h,e} \alpha_n. \tag{4.18}$$

The modal forces $f_m^{h,e}$ can be viewed as discrete counterpart of the modal forces f_m^* introduced in (4.7) and establish the connection between the continuum and the discrete finite element formulation. Comparing (4.18) and (4.7) motivates the introduction of

$$\mathbf{u}_m^{h,e} = \sum_{i=1}^{N} N_i^e(\mathbf{x}) \mathbf{u}_{im}^{*,e} \qquad \text{and} \qquad F_{mn}^{h,e} = \int_{\partial\Omega^e} \mathbf{u}_m^{h,e} \cdot \mathbf{t}_n^* \, \mathrm{d}A. \tag{4.19}$$

In this way, the components $F_{mn}^{h,e}$ have the same structure as the components F_{mn}^* defined in (4.7). In particular, $\mathbf{u}_m^{h,e}$ introduced in (4.19) can be regarded as the finite element alias of the displacement mode $\mathbf{u}_m^*$. We emphasize again that in general both quantities coincide only at the nodes as pointed out in Section 4.2.2.1.

4.2.2.3. Nodal equilibrium

All investigations so far focused on a single finite element Ω^e. However, the correct reproduction of displacement modes $\mathbf{u}_m^*$ on element level is not sufficient to ensure exact solutions for patches of elements. We show subsequently, that it is crucial to additionally fulfill equilibrium of nodal forces.

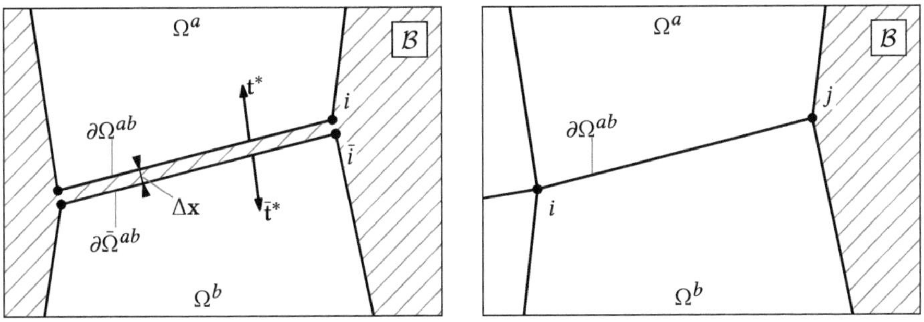

Figure 4.2.: Two finite elements Ω^a and Ω^b with shared boundary $\partial\Omega^{ab}$separated by $\Delta\mathbf{x}$ (left) and a patch of finite elements embedded into the linear elastic continuum (right).

To that end we consider two finite elements Ω^a and Ω^b embedded into the linear elastic continuum $\mathcal{B}$ as shown in Figure 4.2 on the left. Assume that their adjacent faces $\partial\Omega^{ab}$ and $\partial\bar{\Omega}^{ab}$ are congruent and separated by $\Delta\mathbf{x}$. In case of an infinitesimal $\Delta\mathbf{x}$ clearly $\partial\bar{\Omega}^{ab} \longrightarrow \partial\Omega^{ab}$ and $\bar{\mathbf{t}}^* \longrightarrow -\mathbf{t}^*$. Furthermore, the total contribution from boundary $\partial\Omega^{ab}$ to the nodal force at node $\bar{i} \longrightarrow i$ is given by

$$\mathbf{P}^{\mathrm{h}}_{i,ab} := \mathbf{P}^{\mathrm{h},a}_{i,ab} + \mathbf{P}^{\mathrm{h},b}_{i,ab} = \int_{\partial\Omega^{ab}} (N^a_i - N^b_i)\mathbf{t}^* \,\mathrm{d}A, \tag{4.20}$$

where use has been made of (4.17). For nodal equilibrium we require $\mathbf{P}^{\mathrm{h}}_{i,ab} = \mathbf{0}$, which ensures that, after assembly, $\mathbf{P}^{\mathrm{h}}_i = \mathbf{0}$ at every node i inside a finite element patch (see Figure 4.2 on the right). Moreover, contributions from internal faces to nodes at the boundary of an element patch are likewise zero (see, e.g., node j in Figure 4.2 on the right). Thus, if $\mathbf{P}^{\mathrm{h}}_{i,ab} = \mathbf{0}$, all interior contributions vanish and it is possible to consider a complete patch of elements as a single element (with many nodes). Conversely, if nodal equilibrium is fulfilled, it is justified to investigate only a single finite element since results can then be transferred to arbitrary meshes.

To fulfill condition $\mathbf{P}^{\mathrm{h}}_{i,ab} = \mathbf{0}$ for an arbitrary traction $\mathbf{t}^*$, it is obvious in view of (4.20) that a valid choice is given by

$$N^a_i = N^b_i = N_i \qquad \text{on} \qquad \partial\Omega^{ab}. \tag{4.21}$$

With this choice, nodal equilibrium is not only fulfilled for the considered M displacement modes $\mathbf{u}^*_m$ but for arbitrary traction $\mathbf{t}^*$. Condition (4.21) is the classical conformability requirement or, likewise, the C_0 *inter*-element continuity requirement and severely restricts the choice of ansatz functions N_i. Suitable functions can be created by mapping parent elements from the parametric space to the physical space.[8] This approach was indeed introduced to fulfill nodal equilibrium and mapped elements are so successful that they

[8] Isoparametric Lagrangian, serendipity and hierarchical elements as well as the isogeometric approach are examples for such ansatz spaces.

have been used almost exclusively since their first occurrence in the 1960s. All in all, (4.21) is a usual assumption and not as serious of a restriction in practical applications.

Even less restrictive choices to fulfill nodal equilibrium than (4.21) put a restriction on the design of N_i^e. This ultimately establishes MacNeal's theorem as shown in the next section.

Example Q1. *To demonstrate the restrictions imposed by nodal equilibrium we consider again a quadrilateral element: The only sensible choice of parent element for a quadrilateral is a square. All other forms would induce anisotropies into the formulation. Furthermore, the four linearly independent monomials in the parametric space have to be* $1, \xi, \eta, \xi\eta$ *since any other choice would either be incomplete or anisotropic. These restrictions lead directly to the well-known bilinear Lagrangian ansatz functions.*
Thus, there seems to be only one reasonable choice for the ansatz functions of the displacement test function if conformability is to be fulfilled. This illustrates the severe restriction imposed by (4.21).

4.2.3. Design problem

Having summarized properties of the continuum and the finite element formulation allows to finally determine the optimal stiffness matrix for which the finite element is capable to exactly represent $M = N_{\mathrm{DOF}}$ displacement modes. Equation (4.16) together with (4.18) yields the *design problem* [84] of finite element technology

$$\sum_{n=1}^{M} \tilde{K}_{mn}^{\mathrm{h},e} \alpha_n = \sum_{n=1}^{M} F_{mn}^{\mathrm{h},e} \alpha_n, \qquad m = 1, \ldots, M, \tag{4.22}$$

where the goal is to design the element's modal stiffness matrix $\tilde{K}_{mn}^{\mathrm{h},e}$ such that (4.22) holds for arbitrary α_n. If that is the case, all *nodal displacements* $\mathbf{u}_i^e$ would be exact.

Without restrictions (such as (4.21)) on the choice of ansatz functions N_i^e, the design problem would be simple to solve. Proper choice of N_i^e would ensure $F_{mn}^{\mathrm{h},e} = F_{mn}^*$, that is, the finite element would account for the boundary traction in exactly the same way as the continuum solution for the M displacement modes under consideration. In particular, $F_{mn}^{\mathrm{h},e}$ could be diagonalized. It would then be straightforward to choose a proper nodal stiffness matrix $\mathbf{K}^{\mathrm{h},e}$ such that $\tilde{K}_{mn}^{\mathrm{h},e} = F_{mn}^*$.
However, as shown in Section 4.2.2.3, the conformability requirement (4.21) puts severe limitations on the choice of N_i^e. This restriction directly determines $F_{mn}^{\mathrm{h},e}$ through (4.19). Thus, $F_{mn}^{\mathrm{h},e} = F_{mn}^*$ can not be accomplished in general.
By comparing definition (4.19) for $F_{mn}^{\mathrm{h},e}$ with definition (4.7) for F_{mn}^*, we observe that $F_{mn}^{\mathrm{h},e} = F_{mn}^*$ holds in general only if $\mathbf{u}_m^{\mathrm{h},e} = \mathbf{u}_m^*$ on $\partial\Omega^e$. However, as shown in Section 4.2.2.1, the latter equality does not hold in general. More specifically, in case of the widely used *isoparametric concept*, $\mathbf{u}_m^{\mathrm{h},e} = \mathbf{u}_m^*$ can be guaranteed only for linear displacement modes $\mathbf{u}_m^{*,\mathrm{L}}$. In contrast to that, for higher-order displacement modes $\mathbf{u}_m^{*,\mathrm{H}}$, $\mathbf{u}_m^{\mathrm{h},e} \neq \mathbf{u}_m^*$ except for

special cases. Corresponding to these sets of modes $\mathbf{u}_m^{*,\mathrm{L}}$ and $\mathbf{u}_m^{*,\mathrm{H}}$, we may calculate $F_{mn}^{\mathrm{h},e,\mathrm{LL}}$, $F_{mn}^{\mathrm{h},e,\mathrm{HL}}$ and $F_{mn}^{\mathrm{h},e,\mathrm{HH}}$ via (4.19). Now, the right-hand side of (4.22) can be recast in the form

$$[F_{mn}^{\mathrm{h},e}][\alpha_n] = \begin{bmatrix} F_{mn}^{\mathrm{h},e,\mathrm{LL}} & 0 \\ F_{mn}^{\mathrm{h},e,\mathrm{HL}} & F_{mn}^{\mathrm{h},e,\mathrm{HH}} \end{bmatrix} \begin{bmatrix} \alpha_n^{\mathrm{L}} \\ \alpha_n^{\mathrm{H}} \end{bmatrix}, \tag{4.23}$$

where only $F_{mn}^{\mathrm{h},e,\mathrm{LL}}$ is diagonal and equal to the corresponding part of F_{mn}^*. Equation (4.23) clearly shows that $F_{mn}^{\mathrm{h},e}$ is unsymmetric which is induced by the coupling between lower and higher order modes.

Accordingly, satisfaction of design problem (4.22) would require the modal stiffness matrix $\tilde{K}_{mn}^{\mathrm{h},e}$ to be unsymmetric. This in turn would mean an unsymmetric nodal stiffness matrix[9] as can be concluded from (4.15a). Conversely, a symmetric stiffness matrix makes it impossible to simultaneously fulfill the patch test and correctly represent higher-order displacement modes, which is the central statement of MacNeal's theorem. It is worth mentioning that MacNeal's theorem does not rule out special cases such as in the examples below.

Example Q1. *A quadrilateral element can simultaneously satisfy the patch test and exactly represent quadratic displacement modes under certain conditions. An example for this is the EAS element proposed by Simo and Rifai [134]. This element is capable to exactly capture pure bending of a cantilever beam (see Sec. 6.1.1 in [134] and the 3d version of this example in Section 4.5.4). However, the exact response is limited to the special case of rectangular elements. In that case the integrals in (4.7) and (4.19) have the same value and thus* $F_{mn}^{h,e} = F_{mn}^*$ *holds even though* $\mathbf{u}_m^{h,e} \neq \mathbf{u}_m^*$. *For other element shapes the results deteriorate rapidly.*

Example Q2. *Another exemption can be observed for quadratic displacement-based nine-node quadrilaterals (Q2) (see MacNeal [85], Ch. 3.5 and Zienkiewicz et al. [162], Ch. 5.7). These elements are capable of representing element shapes with quadratic borders. However, if we restrict the shape to bilinear forms (i.e., the geometric capabilities of quadrilateral elements), the elements are capable of correctly representing full quadratic polynomials in the physical space. Thus, for bilinear shapes* $\mathbf{u}_m^{h,e} = \mathbf{u}_m^*$ *holds for all quadratic modes and the element is exact for bending problems.*
Higher order modes (cubic and quartic) incorporated in the base of Q2 are unfortunately not represented correctly. Furthermore, it is not advisable to generally restrict the geometry to bilinear shapes in case of curved domains since this degrades the order of convergence with mesh refinement (see Braess [26], Ch. II.§1).

[9] The option of an unsymmetric finite element formulation was already recognized by MacNeal [84]. At that time, however, it seemed "[...] *abhorrent for many reasons* [...]" (MacNeal [84]) to consider such elements. However, previous works [32, 59, 117] and the present contribution show, that there is much to be gained with unsymmetric stiffness matrices.

4.2.4. Summary of MacNeal's Theorem

We summarize the key conclusions of the previous pages in this section. First, a finite element can in general only be "exact" for a maximum of $M = N_{\mathrm{DOF}}$ modes, where N_{DOF} is the number of displacement degrees of freedom. By exact we denote a finite element solution which coincides with the continuum solution at the nodes (nodally exact response). Second, it is essential to fulfill nodal equilibrium to extend the solution capabilities of a single element to a whole patch of elements. This is ensured if the ansatz functions for the test function of the displacement are conforming. Finally, to reach exact representation of a maximum of $M = N_{\mathrm{DOF}}$ displacement modes for general element geometries, the element's stiffness matrix $\mathbf{K}^e$ needs to be unsymmetric. In case of a symmetric $\mathbf{K}^e$ there is always a trade-off between lower and higher order modes (see also MacNeal [84]).

All these conclusions hold for a broad class of finite elements. In particular, they apply to all well-known finite element discretization schemes regardless of their internal design. That is, regardless of whether they are based on mixed formulations, higher-order methods or reduced integration.

4.3. Petrov-Galerkin EAS framework

Our investigations based on MacNeal's theorem presented in the last section indicate that the best performance for a given set of nodes can only be achieved with an unsymmetric stiffness matrix. However, the question on *how* to construct such a stiffness matrix has not been addressed. In order to answer this question, we propose an *Petrov-Galerkin* extension of the *enhanced assumed strain* (EAS) method originally proposed by Simo and Rifai [134]. Our subsequent examinations of this basis yields *three* design conditions that enable the construction of high performance EAS elements.

4.3.1. Weak form

The key idea of the EAS method proposed by Simo and Rifai [134] is to recast the strain in the form $\boldsymbol{\varepsilon} = \nabla^s_{\mathbf{x}}\mathbf{u} + \tilde{\boldsymbol{\varepsilon}}$. Therein $\nabla^s_{\mathbf{x}}\mathbf{u}$ and $\tilde{\boldsymbol{\varepsilon}}$ denote the compatible and incompatible part of the strain, respectively. Inserting this ansatz into the three-field *Hu-Washizu* [148] functional yields the variational framework for the EAS method (see Simo and Rifai [134] for details). In the present work we start with the *discrete* weak form and assume that the independent stress has already been eliminated via a suitable L_2-orthogonality condition. The weak form is then given by (see [134])

$$G^{\mathrm{h}}_{\mathrm{EAS},u} = \int_{\mathcal{B}^h} \nabla^s_{\mathbf{x}}\mathbf{v}^{\mathrm{h}} : \hat{\boldsymbol{\sigma}}^{\mathrm{h}}(\mathbf{u}^{\mathrm{h}}, \tilde{\boldsymbol{\varepsilon}}^{\mathrm{h}})\, \mathrm{d}V - G^{\mathrm{h}}_{\mathrm{ext}}(\mathbf{v}^{\mathrm{h}}) = 0 \qquad \forall\, \mathbf{v}^{\mathrm{h}} \in \mathcal{V}^{\mathrm{h}}, \tag{4.24a}$$

$$G^{\mathrm{h}}_{\mathrm{EAS},\tilde{\varepsilon}} = \int_{\mathcal{B}^h} \tilde{\boldsymbol{\epsilon}}^{\mathrm{h}} : \hat{\boldsymbol{\sigma}}^{\mathrm{h}}(\mathbf{u}^{\mathrm{h}}, \tilde{\boldsymbol{\varepsilon}}^{\mathrm{h}})\, \mathrm{d}V = 0 \qquad \forall\, \tilde{\boldsymbol{\epsilon}}^{\mathrm{h}} \in \mathcal{F}^{\mathrm{h}}, \tag{4.24b}$$

where $\mathbf{v}^{\mathrm{h}} \in \mathcal{V}^{\mathrm{h}}$, $\mathbf{u}^{\mathrm{h}} \in \mathcal{U}^{\mathrm{h}}$ as well as $\tilde{\boldsymbol{\epsilon}}^{\mathrm{h}} \in \mathcal{F}^{\mathrm{h}}$, $\tilde{\boldsymbol{\varepsilon}}^{\mathrm{h}} \in \mathcal{E}^{\mathrm{h}}$ denote the finite element approximations of test and trial functions of the displacement and incompatible strain, respectively. Furthermore,

$$\hat{\boldsymbol{\sigma}}^{\mathrm{h}} = \mathbb{C} : (\nabla_{\mathbf{x}}^{s} \mathbf{u}^{\mathrm{h}} + \tilde{\boldsymbol{\varepsilon}}^{\mathrm{h}}), \tag{4.25}$$

$$G_{\mathrm{ext}}^{\mathrm{h}}(\mathbf{v}^{\mathrm{h}}) = \int_{\mathcal{B}^h} \mathbf{v}^{\mathrm{h}} \cdot \mathbf{b}^{\mathrm{h}} \, \mathrm{d}V + \int_{\partial \mathcal{B}_{\mathrm{t}}^{\mathrm{h}}} \mathbf{v}^{\mathrm{h}} \cdot \bar{\mathbf{t}}^{\mathrm{h}} \, \mathrm{d}A, \tag{4.26}$$

define the constitutive stress $\hat{\boldsymbol{\sigma}}^{\mathrm{h}}$ as well as the contribution of the external forces to the weak form $G_{\mathrm{ext}}^{\mathrm{h}}$ in analogy to G_{ext}^{*} given in (4.3). The external forces $\mathbf{b}^{\mathrm{h}}$ and $\bar{\mathbf{t}}^{\mathrm{h}}$ are the finite element approximations of the continuum counterparts $\mathbf{b}^{*}$ and $\bar{\mathbf{t}}^{*}$. Since the ansatz for the enhanced assumed strain, $\tilde{\boldsymbol{\epsilon}}^{\mathrm{h}} \in \mathcal{F}^{\mathrm{h}}$, $\tilde{\boldsymbol{\varepsilon}}^{\mathrm{h}} \in \mathcal{E}^{\mathrm{h}}$, does not need to be inter-element continuous, it is as usual introduced elementwise, which later allows static condensation of the internal degrees of freedom on element level.

The weak form (4.24) is so far exactly the same as the one presented in Simo and Rifai [134]. The only difference in the present work is that we subsequently apply a *Petrov-Galerkin* approach instead of the usual *Bubnov-Galerkin* method. Thus, we use different ansatz functions for the test and trial spaces. The corresponding ansatz spaces $\mathcal{V}^{\mathrm{h}}, \mathcal{U}^{\mathrm{h}}, \mathcal{F}^{\mathrm{h},e}, \mathcal{E}^{\mathrm{h},e}$ will be specified in Section 4.4.3.

4.3.2. Design conditions for exact nodal solutions

The goal of this section is to find suitable design conditions for EAS elements which ensure that the solution of (4.24) for a given analytic displacement state $\mathbf{u}^{*}$ is nodally exact. That is, the nodal displacements coincide with the analytic displacement state according to (4.12).

The first design condition stems directly from Section 4.2.2.3 where it has been shown that it is crucial to fulfill nodal equilibrium by proper choice of the test function for the displacement $\mathbf{v}^{\mathrm{h},e}$. This allows to consider only a single finite element in analogy to Section 4.2, since results may be generalized to larger patches of elements as shown in Section 4.2.2.3.

To determine further design conditions we start with the analytical weak form for a single finite element given in (4.9) and add the body force contribution, which has been neglected in (4.9). Choosing $\mathbf{v}^{*} = \mathbf{v}^{\mathrm{h},e}$ yields

$$G_{\mathrm{ext}}^{*,e}(\mathbf{v}^{\mathrm{h},e}) = \int_{\Omega^e} \mathbf{v}^{\mathrm{h},e} \cdot \mathbf{b}^{*} \, \mathrm{d}V + \int_{\partial \Omega^e} \mathbf{v}^{\mathrm{h},e} \cdot \mathbf{t}^{*} \, \mathrm{d}A = \int_{\Omega^e} \nabla_{\mathbf{x}}^{s} \mathbf{v}^{\mathrm{h},e} : \boldsymbol{\sigma}^{*} \, \mathrm{d}V. \tag{4.27}$$

Similar to the investigations in Section 4.2.2.2 we set the external loads in (4.26) to $\mathbf{b}^{\mathrm{h}} = \mathbf{b}^{*}$ and $\bar{\mathbf{t}}^{\mathrm{h}} = \mathbf{t}^{*}$. Thus, we obtain identity $G_{\mathrm{ext}}^{*,e}(\mathbf{v}^{\mathrm{h},e}) = G_{\mathrm{ext}}^{\mathrm{h},e}(\mathbf{v}^{\mathrm{h},e})$. Finally substituting this result into (4.24a) and enforcing the weak form for a single finite element gives

$$\boxed{\int_{\Omega^e} \nabla_{\mathbf{x}}^{s} \mathbf{v}^{\mathrm{h},e} : \left(\hat{\boldsymbol{\sigma}}^{\mathrm{h},e} - \boldsymbol{\sigma}^{*}\right) \, \mathrm{d}V = 0} \tag{4.28}$$

which determines the next condition for an exact finite element solution. Since we have seen that nodal equilibrium puts severe restrictions on the choice of the test functions $\mathbf{v}^{h,e}$ (see also the first design condition), it is in general only possible to fulfill (4.28) if $\hat{\boldsymbol{\sigma}}^{h,e} = \boldsymbol{\sigma}^*$ holds pointwise, where, according to (4.1b), $\boldsymbol{\sigma}^* = \mathbb{C} : \nabla^s_{\mathbf{x}} \mathbf{u}^*$. On the other hand, the constitutive stress (4.25) is given by $\hat{\boldsymbol{\sigma}}^{h,e} = \mathbb{C} : (\nabla^s_{\mathbf{x}} \mathbf{u}^{h,e} + \tilde{\boldsymbol{\varepsilon}}^{h,e})$. Thus, condition $\hat{\boldsymbol{\sigma}}^{h,e} = \boldsymbol{\sigma}^*$ can be fulfilled by choosing the ansatz spaces for $\mathbf{u}^{h,e}$ and $\tilde{\boldsymbol{\varepsilon}}^{h,e}$ appropriately. Furthermore, $\hat{\boldsymbol{\sigma}}^{h,e} = \boldsymbol{\sigma}^*$ must be possible under the premise that the nodal displacements are exact. Otherwise it would be possible to fulfill (4.28) without having exact nodal displacements.

The third and final condition emerges from the second equation of the weak form. Substituting $\hat{\boldsymbol{\sigma}}^{h,e} = \boldsymbol{\sigma}^*$ (i.e., the second condition) into (4.24b) yields

$$\boxed{\int_{\Omega^e} \tilde{\boldsymbol{\epsilon}}^{h,e} : \boldsymbol{\sigma}^* \, dV = 0,} \tag{4.29}$$

which determines that the test functions of the incompatible strain has to be L_2-orthogonal to the stress $\boldsymbol{\sigma}^*$. Interestingly, this is an extended version of the patch-test condition of the classical EAS element by Simo and Rifai [134]. However, instead of requiring (4.29) only for constant stress fields we demand L_2-orthogonality for all modes for which the element should be exact.

Summarizing the above we established three conditions, which have to be fulfilled in order to construct a Petrov-Galerkin EAS element on the basis of (4.24) that is exact for a chosen displacement field $\mathbf{u}^*$ regardless of element shape:

- **C1** The test functions $\mathbf{v}^{h,e} \in \mathcal{V}^h$ have to ensure that nodal equilibrium is fulfilled regardless of element geometry and neighboring elements (see Section 4.2.2.3),
- **C2** the discrete constitutive stress tensor $\hat{\boldsymbol{\sigma}}^{h,e}$, which is computed from $\mathbf{u}^{h,e} \in \mathcal{U}^h$ and $\tilde{\boldsymbol{\varepsilon}}^{h,e} \in \mathcal{E}^h$ via (4.25), must include $\boldsymbol{\sigma}^*$ under the premise that the nodal displacements are exact (see (4.28)) and
- **C3** the test functions for the enhanced strain $\tilde{\boldsymbol{\epsilon}}^{h,e} \in \mathcal{F}^h$ must be L_2-orthogonal to $\boldsymbol{\sigma}^*$ (see (4.29)).

4.4. Design of mesh distortion insensitive EAS elements

The final theoretical part of this contribution concerns the actual finite element design. We have to specify the analytic modes, for which the element should be exact, and then construct the ansatz spaces for all four fields introduced in (4.24) such that Conditions C1 to C3 are fulfilled. We start with the introduction of the skew coordinate frame in Section 4.4.1 since it is the key for the analytic stress modes in Section 4.4.2 and the finite element discretization in Section 4.4.3. For simplicity, we focus on the 2D case in Sections 4.4.1 to 4.4.3 and cover the extension to 3D in Section 4.4.4.

4.4.1. Skew coordinate frame

An element Ω^e is described as usual by mapping a reference element $\hat{\Omega} = [-1, +1]^2$ to $\Omega^e \subset \mathbb{R}^2$. Accordingly, each point with parametric coordinates $\boldsymbol{\xi} = \begin{bmatrix}\xi & \eta\end{bmatrix}^\mathrm{T} \in \hat{\Omega}$ is mapped to a physical point $\mathbf{x} = \begin{bmatrix}x & y\end{bmatrix}^\mathrm{T} \in \Omega^e$ via the *isoparametric* transformation (see Figure 4.3)

$$\mathbf{x}^{\mathrm{h},e} = \sum_{i=1}^{4} N_i(\boldsymbol{\xi})\mathbf{x}_i^e, \qquad \mathbf{J}^{\mathrm{h},e} = \frac{\partial \mathbf{x}^{\mathrm{h},e}}{\partial \boldsymbol{\xi}} = \sum_{i=1}^{4} \mathbf{x}_i^e \otimes \nabla_{\boldsymbol{\xi}} N_i. \tag{4.30}$$

Therein, $\mathbf{J}^{\mathrm{h},e}$ is the Jacobian of the transformation $\mathbf{x}^{\mathrm{h},e} : \hat{\Omega} \mapsto \Omega^e$, $\mathbf{x}_i^e$ are the nodal coordinates and

$$N_i = \frac{1}{4}(1 - \xi_i\xi)(1 - \eta_i\eta) \tag{4.31}$$

denote the usual bi-linear Lagrangian shape functions. They are defined with help of the nodal positions $\begin{bmatrix}\xi_i & \eta_i\end{bmatrix}^\mathrm{T} = \begin{bmatrix}\pm 1 & \pm 1\end{bmatrix}^\mathrm{T}$ in the parametric space.

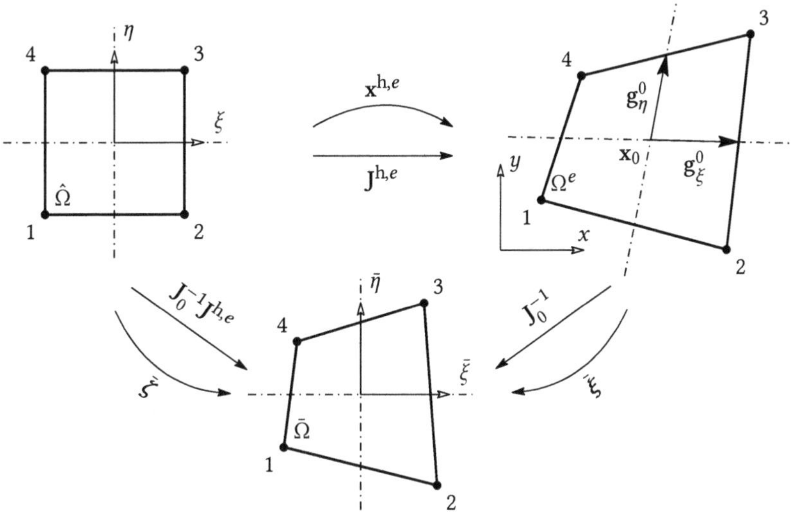

Figure 4.3.: Geometry maps and coordinate systems for the four node quadrilateral element. Isoparametric map $\mathbf{x}^{\mathrm{h},e} : \hat{\Omega} \mapsto \Omega^e$ (see (4.30)), skew map $\bar{\boldsymbol{\xi}} : \Omega^e \mapsto \bar{\Omega}$ (see (4.32)) and additional map $\bar{\boldsymbol{\zeta}} : \hat{\Omega} \mapsto \bar{\Omega}$ (see (4.35)).

In the context of assumed stress elements Yuan et al. [159] and Wisniewski and Turska [150] proposed to use a *skew coordinate frame* for an alternative description of geometry. The skew coordinates are introduced via the elementwise affine map $\bar{\boldsymbol{\xi}} : \Omega^e \mapsto \bar{\Omega}$ defined by

$$\bar{\boldsymbol{\xi}} = \begin{bmatrix}\bar{\xi} & \bar{\eta}\end{bmatrix}^\mathrm{T} = \mathbf{J}_0^{-1}(\mathbf{x}^{\mathrm{h},e} - \mathbf{x}_0), \tag{4.32}$$

where $\mathbf{J}_0 = \mathbf{J}^{\mathrm{h},e}\big|_{\boldsymbol{\xi}=\mathbf{0}}$ is the Jacobian at the element center $\mathbf{x}_0 = \mathbf{x}^{\mathrm{h},e}\big|_{\boldsymbol{\xi}=\mathbf{0}}$. The inverse of transformation (4.32) is given by

$$\mathbf{x}^{\mathrm{h},e} = \mathbf{J}_0 \bar{\boldsymbol{\xi}} + \mathbf{x}_0. \tag{4.33}$$

Since the covariant base vectors $\mathbf{g}_\xi^0 = \partial \mathbf{x}^{\mathrm{h},e}/\partial\xi\big|_{\boldsymbol{\xi}=\mathbf{0}}$ and $\mathbf{g}_\eta^0 = \partial \mathbf{x}^{\mathrm{h},e}/\partial\eta\big|_{\boldsymbol{\xi}=\mathbf{0}}$ constitute the columns of $\mathbf{J}_0$, (4.33) can also be written as $\mathbf{x}^{\mathrm{h},e} = \bar{\xi}\mathbf{g}_\xi^0 + \bar{\eta}\mathbf{g}_\eta^0 + \mathbf{x}_0$. Accordingly, the coordinates $\bar{\xi}$ and $\bar{\eta}$ refer to the base vectors $\mathbf{g}_\xi^0$ and $\mathbf{g}_\eta^0$, which in general span a skew basis (see Figure 4.3). It can further be observed from Figure 4.3 that the skew element $\bar{\Omega}$ plays the role of an intermediate configuration between the reference element $\hat{\Omega}$ and the physical element Ω^e.

Two important properties of the skew coordinates follow directly from (4.32). First, since the map (4.32) from $\bar{\Omega}$ to Ω^e is affine, the two elements $\bar{\Omega}$ and Ω^e are affine-equivalent (see, e.g., Ch. 12 in Reddy [119]). This implies in particular that complete polynomials defined on $\bar{\Omega}$ are mapped to complete polynomials in Ω^e. This property will be exploited below for the construction of complete ansatz functions regardless of the geometry of element Ω^e. Second, the skew coordinates are frame-indifferent. This can be shown by considering superposed rigid motions corresponding to nodal position changes of the form $\mathbf{x}_i^{e\#} = \mathbf{R}\mathbf{x}_i^e + \mathbf{c}$, where $\mathbf{R} \in SO(2)$ is a rotation tensor and $\mathbf{c} \in \mathbb{R}^2$ is an arbitrary vector. Substituting $\mathbf{x}_i^{e\#}$ for $\mathbf{x}_i^e$ in (4.30) yields $\mathbf{x}^{\mathrm{h},e\#} = \mathbf{R}\mathbf{x}^{\mathrm{h},e} + \mathbf{c}$ and $\mathbf{J}^{\mathrm{h},e\#} = \mathbf{R}\mathbf{J}^{\mathrm{h},e}$. With regard to (4.32), we then obtain

$$\bar{\boldsymbol{\xi}}^{\#} = \mathbf{J}_0^{\#^{-1}}(\mathbf{x}^{\mathrm{h},e\#} - \mathbf{x}_0^{\#}) = (\mathbf{R}\mathbf{J}_0)^{-1}\mathbf{R}(\mathbf{x}^{\mathrm{h},e} - \mathbf{x}_0) = \mathbf{J}_0^{-1}\mathbf{R}^{-1}\mathbf{R}(\mathbf{x}^{\mathrm{h},e} - \mathbf{x}_0) = \bar{\boldsymbol{\xi}}, \tag{4.34}$$

which establishes the frame-indifference of the skew coordinates. This facilitates subsequently the construction of a frame-indifferent finite element formulation.

Remark 4.3. *Although* (4.32) *links the skew coordinates to the parametric coordinates via* $\mathbf{x}^{h,e}$ *and* (4.30), *it is convenient to have a direct relation between* $\boldsymbol{\xi}$ *and* $\bar{\boldsymbol{\xi}}$. *In particular, the map* $\bar{\boldsymbol{\xi}}: \hat{\Omega} \mapsto \bar{\Omega}$ *is given by*

$$\bar{\boldsymbol{\xi}} = \boldsymbol{\xi} + \mathbf{J}_0^{-1}\mathbf{c}_1 H_1(\boldsymbol{\xi}), \qquad \mathbf{c}_1 = \frac{1}{4}\sum_{i=1}^{4} \mathbf{x}_i^e h_i^1, \tag{4.35}$$

where $H_1 = \xi\eta$ *and* $\mathbf{h}^1 = \begin{bmatrix} +1 & -1 & +1 & -1 \end{bmatrix}^{\mathrm{T}}$ *are the hourglass function and vector, respectively (see Appendix 4.B for further details). It is straightforward to verify that* $\mathbf{c}_1 = \mathbf{0}$ *for a parallelogram shaped element and therefore* $\bar{\boldsymbol{\xi}} = \boldsymbol{\xi}$ *in case of a constant Jacobian.*

4.4.2. Prescribed analytic modes

The investigations of Section 4.2.2.2 show that the four node quadrilateral element with eight degrees of freedom can be fitted to a maximum of $M = N_{\mathrm{DOF}} = 8$ prescribed modes. Three of which are necessarily the rigid body modes, since elements that are not exact for those would lead to completely unphysical results. For the remaining five modes we propose to use the *assumed stress* modes developed by Pian and Sumihara [110]. These modes are suited perfectly since they include the important constant stress modes as well as the bending modes. The former are necessary to fulfill the patch test [162] and the latter are of the utmost importance in many engineering applications. Furthermore, the almost unchanged form of the assumed stress modes for more than 35 years suggests their optimality for assumed stress elements.

One of the few successful modifications of the Pian-Sumihara stress modes is formulating them in the skew coordinate frame. This has the advantage that equilibrium, that is, $\operatorname{div} \boldsymbol{\sigma}^{\mathrm{h},e} = \mathbf{0}$, is fulfilled a priori regardless of element shape [150, 159]. We therefore choose the analytic stress modes in the skew system, which are in vector matrix form given by

$$\bar{\sigma}^*_v(\bar{\xi}) := \begin{bmatrix} 1 & 0 & 0 & \bar{\eta} & 0 \\ 0 & 1 & 0 & 0 & \bar{\xi} \\ 0 & 0 & 1 & 0 & 0 \end{bmatrix} \begin{bmatrix} \beta_1 \\ \vdots \\ \beta_5 \end{bmatrix}, \tag{4.36}$$

as prescribed analytic modes. These stress modes have to be transformed to the physical frame to use them in (4.28). The usual assumption of a contravariant stress field yields the transformation

$$\boldsymbol{\sigma}^* = \mathbf{J}_0 \bar{\boldsymbol{\sigma}}^* \mathbf{J}_0^{\mathrm{T}}. \tag{4.37}$$

The last two equations could be used to compute the corresponding displacement modes $\mathbf{u}^*_m$, $m = 1, \dots, 5$ via the system of coupled PDEs (4.1). Fortunately, it is not necessary to perform this tedious task. To fulfill conditions (4.28) and (4.29) we only need the prescribed stress $\boldsymbol{\sigma}^*$. By proper choice of the ansatz space of the displacement and the incompatible strain we can ensure that the nodal displacements are exact without knowing the actual value of the analytic displacement at the node.

4.4.3. Ansatz spaces

The final task is to find ansatz spaces that fulfill Conditions C1 to C3 for the prescribed modes (4.37) such that we obtain an element that is nodally exact for these deformation states.

4.4.3.1. Test function for the displacement

Condition C1, which concerns nodal equilibrium, is straightforward to fulfill. The only requirement is a conforming ansatz (4.21) for the test functions of the displacement $\mathbf{v}^{\mathrm{h},e}$. Since this is exactly what the *isoparametric* concept was designed for, we choose

$$\mathbf{v}^{\mathrm{h},e} = \sum_{i=1}^{4} N_i \mathbf{v}_i^e \tag{4.38}$$

which uses the same bi-linear shape functions (4.31) as the ansatz for the geometry (4.30).

4.4.3.2. Trial function for the displacement

Condition C2 determines that the discrete stress $\hat{\boldsymbol{\sigma}}^{\mathrm{h},e}$ has to include the prescribed stress modes $\boldsymbol{\sigma}^*$ given in (4.37). Since the chosen stress modes (4.36) are linear with respect to the skew coordinates, we meet Condition C2 if each component of $\nabla_{\mathbf{x}}^s \mathbf{u}^{\mathrm{h},e} + \tilde{\boldsymbol{\varepsilon}}^{\mathrm{h},e}$ contains complete linear polynomials $P_1(\Omega^e)$. Then, in view of the linear map (4.25), the discrete constitutive stress $\hat{\boldsymbol{\sigma}}^{\mathrm{h},e}$ is also a complete linear polynomial and contains in particular $\boldsymbol{\sigma}^*$. This approach is more general than explicitly needed for Condition C2, since we can represent not only (4.36) but complete linear stress modes. However, its independence of the material model (i.e., $\mathbb{C}$) is a crucial advantage.

Although the isoparametric concept ensures that complete linear displacement modes are correctly represented in the physical space, this is in general not the case for higher-order displacement modes. This deficiency is caused by the nonlinearity of the map (4.30).

Example Q1. *The components of the map* (4.30) *can be written as* $x = a_0 + a_1\xi + a_2\eta + a_3\xi\eta$ *and* $y = b_0 + b_1\xi + b_2\eta + b_3\xi\eta$ *with constants* a_i, b_i. *Unless* $a_3 = b_3 = 0$ *(a parallelogram) the expression* xy *contains monomials up to* $\xi^2\eta^2$ *not included in the base of the isoparametric interpolation. The expression* xy *can therefore not be represented by the bilinear Lagrangian shape functions. Similar relations can be found for all isoparametric elements.*

Complete polynomials in the physical space can be obtained by using shape functions constructed in the physical space, which are termed *metric* shape functions (see the book by MacNeal [85][10]). The construction of metric shape functions starts by choosing an appropriate set of N monomials. For the four node quadrilateral ($N = 4$) the only sensible choice is $1, x, y, xy$. Furthermore, it is necessary to construct the shape functions using the skew coordinates $\bar{\boldsymbol{\xi}}$ instead of the physical coordinates $\mathbf{x}$ to maintain frame-indifference (see Xie et al. [158]). Despite using the skew frame, the shape functions can still be

[10] MacNeal [85] uses the expression metric interpolations instead of metric shape functions.

regarded as metric shape functions since the two elements $\bar{\Omega}$ and Ω^e are affine-equivalent (see Section 4.4.1). Following these considerations, we choose

$$u^{\mathrm{h},e}(\bar{\xi}) = \begin{bmatrix} 1 & \bar{\xi} & \bar{\eta} & \bar{\xi}\bar{\eta} \end{bmatrix} \begin{bmatrix} a_1 \\ \vdots \\ a_4 \end{bmatrix} \tag{4.39}$$

as ansatz for the x-component of the displacement vector. The modal weights a_i, $i = 1, \dots, 4$ can be linked to the nodal displacement weights u_i^e by imposing conditions $u^{\mathrm{h},e}(\bar{\xi}_i) = u_i^e$, $i = 1, \dots, 4$, where $\bar{\xi}_i$ denote the skew coordinates of the nodes of element $\bar{\Omega}$. Doing so yields the metric shape functions $M_i^e(\bar{\xi})$ defined through

$$u^{\mathrm{h},e}(\bar{\xi}) = \begin{bmatrix} 1 & \bar{\xi} & \bar{\eta} & \bar{\xi}\bar{\eta} \end{bmatrix} \begin{bmatrix} 1 & \bar{\xi}_1 & \bar{\eta}_1 & \bar{\xi}_1\bar{\eta}_1 \\ & & \vdots & \\ 1 & \bar{\xi}_4 & \bar{\eta}_4 & \bar{\xi}_4\bar{\eta}_4 \end{bmatrix}^{-1} \begin{bmatrix} u_1^e \\ \vdots \\ u_4^e \end{bmatrix} = \begin{bmatrix} M_1^e & \cdots & M_4^e \end{bmatrix} \begin{bmatrix} u_1^e \\ \vdots \\ u_4^e \end{bmatrix}. \tag{4.40}$$

The inverse matrix required to compute M_i^e exists as long as there are no coincident nodes. It is also well-conditioned even for elements with high aspect ratios because of the skew coordinates which scale the element appropriately. Naturally, both displacement components of $\mathbf{u}^{\mathrm{h},e}$ have to be approximated using the same M_i^e to maintain isotropy. The final approximation of the trial function for the displacement reads

$$\mathbf{u}(\bar{\xi})^{\mathrm{h},e} = \sum_{i=1}^{4} M_i^e(\bar{\xi})\mathbf{u}_i^e \tag{4.41}$$

in analogy to (4.38). Starting from (4.40) it is now straightforward to compute the derivative of the displacement component $u^{\mathrm{h},e}$ with respect to $\bar{\xi}$ which is given by

$$\nabla_{\bar{\xi}} u^{\mathrm{h},e} = \begin{bmatrix} 0 & 1 & 0 & \bar{\eta} \\ 0 & 0 & 1 & \bar{\xi} \end{bmatrix} \begin{bmatrix} 1 & \bar{\xi}_1 & \bar{\eta}_1 & \bar{\xi}_1\bar{\eta}_1 \\ & & \vdots & \\ 1 & \bar{\xi}_4 & \bar{\eta}_4 & \bar{\xi}_4\bar{\eta}_4 \end{bmatrix}^{-1} \begin{bmatrix} u_1^e \\ \vdots \\ u_4^e \end{bmatrix} = \begin{bmatrix} \nabla_{\bar{\xi}} M_1^e & \cdots & \nabla_{\bar{\xi}} M_4^e \end{bmatrix} \begin{bmatrix} u_1^e \\ \vdots \\ u_4^e \end{bmatrix}. \tag{4.42}$$

The last equation yields expressions for the derivatives of the metric shape functions $\nabla_{\bar{\xi}} M_i^e$, $i = 1, \dots, 4$. Derivatives with respect to the physical coordinates follow from application of the chain rule together with (4.32) such that

$$\nabla_{\mathbf{x}} M_i^e = \mathbf{J}_0^{-\mathrm{T}} \nabla_{\bar{\xi}} M_i^e. \tag{4.43}$$

The derivatives $\nabla_{\bar{\xi}} M_i^e$ are obviously not complete linear polynomials which prohibits fulfilling Condition C2 using only the displacement ansatz (4.41). It is further worth noting that the inverse matrices in (4.40) and (4.42) required to calculate the metric shape functions M_i^e and their derivatives $\nabla_{\bar{\xi}} M_i^e$ are the same and need to be computed only once per element.

We summarize a few properties of the metric ansatz described above which are straightforward to verify using (4.40) before we conclude this section. Functions M_i^e are by construction frame-indifferent and isotropic [158]. The latter implies that M_i^e are invariant with respect to the node-numbering, that is, the direction of the skew basis vectors $\mathbf{g}_i^0$. Furthermore, they are a partition of unity $\sum M_i^e = 1$ and have the Kronecker-delta property $M_i^e(\mathbf{x}_j) = \delta_{ij}$. Finally, M_i^e are equal to the bilinear shape functions N_i in case of elements with constant Jacobian $\mathbf{J}^{\mathrm{h},e}$. However, M_i^e are *non-conforming* for general element shapes, that is (4.21) does not hold as shown in Figure 4.4.

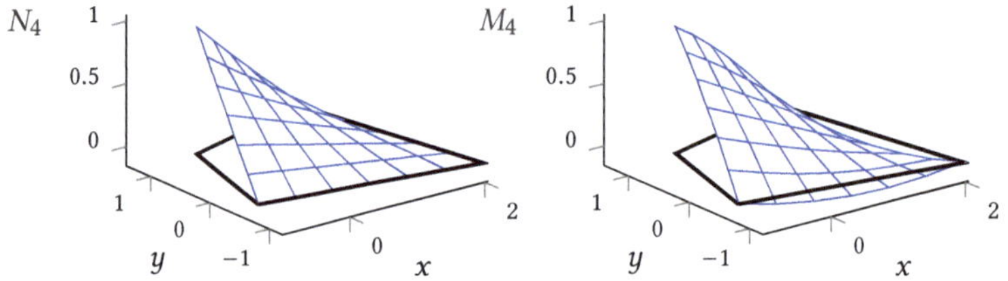

Figure 4.4.: Isoparametric shape function N_4 (left) and corresponding metric shape function M_4 (right) of a quadrilateral finite element Ω^e with nodes $\mathbf{x}_1 = [-1, -0.8]$, $\mathbf{x}_2 = [2, -1.2]$, $\mathbf{x}_3 = [1.5, 1.3]$, $\mathbf{x}_4 = [-0.5, 0.8]$.

4.4.3.3. Trial function for the enhanced strain

Relation (4.42) reveals that the derivatives of the metric shape functions are not complete linear polynomials and have to be complemented in order to fulfill Condition C2. Inspection of (4.40) shows that the missing expressions for a complete quadratic displacement field are $\tilde{m}_1^e = \bar{\xi}^2$ and $\tilde{m}_2^e = \bar{\eta}^2$. A simple way to add these terms is considering two elementwise *incompatible mode* ansatz functions with internal weights $\tilde{u}_j^e$. In this way we obtain a displacement field complete in $P_2(\bar{\Omega})$ given by

$$u^{\mathrm{h},e}(\bar{\xi}) = \sum_{i=1}^{4} M_i^e(\bar{\xi}) u_i^e + \sum_{j=1}^{2} \tilde{M}_j^e(\bar{\xi}) \tilde{u}_j^e. \tag{4.44}$$

However, directly choosing $\tilde{M}_j^e = \tilde{m}_j^e$ does not fulfill the second part of Condition C2 which states that identity $\hat{\boldsymbol{\sigma}}^{\mathrm{h},e} = \boldsymbol{\sigma}^*$ must hold under the premise that the nodal displacements u_i^e associated with M_i^e are exact. In other words, the incompatible modes may not contribute to the displacement at the nodes. Choosing

$$\tilde{M}_j^e(\bar{\xi}) = \tilde{m}_j^e(\bar{\xi}) - \sum_{i=1}^{4} M_i^e(\bar{\xi})\, \tilde{m}_j^e(\bar{\xi}_i). \tag{4.45}$$

as ansatz in (4.44) ensures $\tilde{M}_j^e = 0$ at the element's nodes while maintaining the quadratic completeness of the displacement field (4.44).

It is now straightforward to obtain a suitable ansatz for the trial function of the incompatible strain from (4.44) using the strain-displacement relationship of the form (4.1a). Accordingly, taking the derivative of the second term on the right-hand side of (4.44) results in

$$\tilde{\boldsymbol{\varepsilon}}^{\mathrm{h},e} = \sum_{j=1}^{2} \tilde{\mathbf{u}}_j^e \otimes \nabla_{\mathbf{x}}^s \tilde{M}_j^e, \tag{4.46}$$

where the incompatible displacement vectors $\tilde{\mathbf{u}}_j^e$ contain the four internal degrees of freedom.

Remark 4.4. *The shape functions $\tilde{M}_j^e$ presented above are a more compact form of the ansatz proposed by Huang and Li [57] in a Bubnov-Galerkin frame. Furthermore, a similar idea of using metric incompatible modes has recently been suggested by Huang et al. [59]. Their idea is to employ a higher order parent element which automatically ensures $\tilde{M}_j^e = 0$ at the nodes. The drawbacks of that method are the higher number of internal degrees of freedom (at least twice as much), the higher-order of Gauss quadrature required and the difficulty to extend it to 3D problems due to a violation of Condition C3. All of these drawbacks are overcome with the present approach.*

4.4.3.4. Test function for the enhanced strain

The remaining task is to find appropriate test functions for the incompatible strain $\tilde{\boldsymbol{\epsilon}}^{\mathrm{h},e}$ such that Condition C3 holds. To that end, we start with the incompatible strain field of the classical EAS approach [134]

$$\tilde{\mathbf{e}}^{\mathrm{h},e} = \frac{1}{j^{\mathrm{h},e}} \mathbf{J}_0^{-\mathrm{T}} \bar{\mathbf{e}} \mathbf{J}_0^{-1}, \qquad \bar{\mathbf{e}}_v := \begin{bmatrix} \xi & 0 & 0 & 0 \\ 0 & \eta & 0 & 0 \\ 0 & 0 & \xi & \eta \end{bmatrix} \begin{bmatrix} \alpha_1 \\ \vdots \\ \alpha_4 \end{bmatrix}, \tag{4.47}$$

where $j^{\mathrm{h},e} = \det \mathbf{J}^{\mathrm{h},e}$ and $\bar{\mathbf{e}}_v$ contains the classical *Wilson-modes* in vector-matrix notation. The transformations chosen in (4.47) and (4.37) allow to recast Condition C3, that is, (4.29) in the form

$$\int_{\hat{\Omega}} \bar{\mathbf{e}} : \bar{\sigma}^* \, \mathrm{d}\hat{\Omega} = 0. \tag{4.48}$$

This condition is fulfilled in case of the special choice (4.47). Thus Condition C3 is immediately fulfilled by choosing $\tilde{\boldsymbol{\epsilon}}^{\mathrm{h},e} = \tilde{\mathbf{e}}^{\mathrm{h},e}$. However, in case of general ansatz functions, this is not necessarily the case. See, e.g., the 3D case described in Section 4.4.4.3.

Remark 4.5. *Interestingly,* (4.48) *offers an explanation why transformation* (4.47), *which works well in linear simulations [134], is the best choice of transformation of the enhanced modes in nonlinear simulations even though it is not the obvious choice if the deformation*

gradient is enhanced (see Pfefferkorn and Betsch [104]): Since any nonlinear element should be close to the linear case for moderate deformations we may conclude that (4.48) *is required for good bending accuracy even in nonlinear simulations. In case of rectangular elements* (4.48) *is fulfilled for the standard (transposed) Wilson ansatz functions if transformation* (4.47) *is used. This holds only approximately for distorted meshes but still improves performance. Other transformations lead to more severe violations of* (4.48) *which implies worse overall performance.*
This effect is described in numerous works (e.g., [47, 64, 104]) with numerical investigations but could, to the best knowledge of the authors, not be explained so far.

4.4.3.5. Additional features

Exact numerical integration
Similar to other Petrov-Galerkin finite elements [116] standard Gauss quadrature allows to evaluate the integrals in the weak form (4.24) exactly. To show this we first note that the constitutive stress $\hat{\boldsymbol{\sigma}}^{\mathrm{h},e}$ contains only polynomials of x, y since we choose metric interpolations of $\mathbf{u}^{\mathrm{h},e}$ and $\tilde{\boldsymbol{\varepsilon}}^{\mathrm{h},e}$. In view of (4.35), these expressions are also simple polynomials of ξ, η. Next, transformation of the integral in (4.24b) to the reference element is possible using $\mathrm{d}V = j^{\mathrm{h},e}\,\mathrm{d}\hat{\Omega}$. Due to the specific transformation in (4.47) $j^{\mathrm{h},e}$ immediately cancels out which establishes the polynomial form of the integrand in (4.24b). The same holds for (4.24a). Here we can recast the gradient of the isoparametric shape functions in the form

$$\nabla_{\mathbf{x}} N_i = (\mathbf{J}^{\mathrm{h},e})^{-\mathrm{T}} \nabla_{\boldsymbol{\xi}} N_i = \frac{1}{j^{\mathrm{h},e}} \operatorname{cof}(\mathbf{J}^{\mathrm{h},e}) \nabla_{\boldsymbol{\xi}} N_i, \tag{4.49}$$

where $\operatorname{cof}(\bullet)$ denotes the cofactor [24]. This shows that $j^{\mathrm{h},e}$ cancels analogously after transformation of the integral in (4.24a) and the remaining expression again contains only polynomials of ξ, η. By inspection of the integrands we find that the highest order monomials are $\xi^2\eta$, $\xi^2\eta^2$ and $\xi\eta^2$, all of which are integrated correctly by standard 2×2 Gauss quadrature. Thus, standard integration is exact for the present element regardless of element shape.

Connection to the standard EAS element
If an element is parallelogram shaped, the present approach reduces to the standard EAS method presented in the work of Simo and Rifai [134]. In that case the Jacobian of the isoparmetric map $\mathbf{J}^{\mathrm{h},e} = \mathbf{J}_0$ is constant and (4.35) implies $\bar{\boldsymbol{\xi}} = \boldsymbol{\xi}$. We then obtain from (4.40) that $M_i^e = N_i$. Finally, (4.45) become the classical Wilson-bubble modes which establishes the equivalence.

Eliminated orthogonal stress field

The weak form introduced in (4.24) is a reduced form of the full Hu-Washizu framework as mentioned in Section 4.3.1. The complete weak form on element level is given by (see [134])

$$\int_{\mathcal{B}^{\mathrm{h}}} \nabla_{\mathbf{x}}^{s} \mathbf{v}^{\mathrm{h}} : \hat{\boldsymbol{\sigma}}^{\mathrm{h}} \, \mathrm{d}V - G_{\mathrm{ext}}^{\mathrm{h}}(\mathbf{v}^{\mathrm{h}}) = 0 \qquad \forall\, \mathbf{v}^{\mathrm{h}} \in \mathcal{V}^{\mathrm{h}}, \tag{4.50a}$$

$$\int_{\mathcal{B}^{\mathrm{h}}} \tilde{\boldsymbol{\epsilon}}^{\mathrm{h}} : \left(\hat{\boldsymbol{\sigma}}^{\mathrm{h}} - \boldsymbol{\sigma}^{\mathrm{h}}\right) \mathrm{d}V = 0 \qquad \forall\, \tilde{\boldsymbol{\epsilon}}^{\mathrm{h}} \in \mathcal{F}^{\mathrm{h}}, \tag{4.50b}$$

$$\int_{\mathcal{B}^{\mathrm{h}}} \boldsymbol{\tau}^{\mathrm{h}} : \tilde{\boldsymbol{\varepsilon}}^{\mathrm{h}} \, \mathrm{d}V = 0 \qquad \forall\, \boldsymbol{\tau}^{\mathrm{h}} \in \mathcal{T}^{\mathrm{h},e}, \tag{4.50c}$$

where in contrast to (4.24) the stress field $\boldsymbol{\sigma}^{\mathrm{h}}$ with its test functions $\boldsymbol{\tau}^{\mathrm{h}}$ has not yet been eliminated via L_2-orthogonality condition. It is straightforward to get from (4.50) to (4.24). Simply choosing $\boldsymbol{\sigma}^{\mathrm{h},e} = \boldsymbol{\sigma}^*$ for the independent stress field on element level, where $\boldsymbol{\sigma}^*$ is given by (4.37) and (4.36), eliminates $\boldsymbol{\sigma}^{\mathrm{h}}$ from (4.50b) due to (4.29). Moreover, it is simple to construct an ansatz for the test functions for the independent stress $\boldsymbol{\tau}^{\mathrm{h}}$ such that (4.50c) is fulfilled by construction. To that end we simply start by taking (4.37) and replace the skew coordinates $\bar{\boldsymbol{\xi}}$ in (4.36) with the parametric coordinates $\boldsymbol{\xi}$. This set of ansatz functions can then be orthogonalized to $\tilde{\boldsymbol{\varepsilon}}^{\mathrm{h},e}$ with a Gram-Schmidt orthogonalization similar to Section 4.4.4.2. Thus, it is possible to use (4.24) instead of (4.50).

4.4.4. Three dimensional problems

So far, we described the 2D case in a detailed way. However, most of the ideas and methods presented are straightforward to transfer to the 3D setting. We list differences and specifics for the 3D case in this section.

4.4.4.1. Prescribed stress modes

Similar to Section 4.4.2 we use the stress modes from assumed stress elements as prescribed analytic modes. They were first proposed in the work of Pian and Tong [112] in the parametric space. In skew coordinates (4.32), the assumed stress modes are defined by using (4.37) together with

$$\bar{\sigma}_v^*(\bar{\xi}) := \left[\begin{array}{cccccc|ccccccccc|ccc}
1 & 0 & 0 & 0 & 0 & 0 & \bar{\eta} & \bar{\zeta} & 0 & 0 & 0 & 0 & 0 & 0 & 0 & \bar{\eta}\bar{\zeta} & 0 & 0 \\
0 & 1 & 0 & 0 & 0 & 0 & 0 & 0 & \bar{\zeta} & \bar{\xi} & 0 & 0 & 0 & 0 & 0 & 0 & \bar{\zeta}\bar{\xi} & 0 \\
0 & 0 & 1 & 0 & 0 & 0 & 0 & 0 & 0 & 0 & \bar{\xi} & \bar{\eta} & 0 & 0 & 0 & 0 & 0 & \bar{\xi}\bar{\eta} \\
0 & 0 & 0 & 1 & 0 & 0 & 0 & 0 & 0 & 0 & 0 & 0 & \bar{\zeta} & 0 & 0 & 0 & 0 & 0 \\
0 & 0 & 0 & 0 & 1 & 0 & 0 & 0 & 0 & 0 & 0 & 0 & 0 & \bar{\xi} & 0 & 0 & 0 & 0 \\
0 & 0 & 0 & 0 & 0 & 1 & 0 & 0 & 0 & 0 & 0 & 0 & 0 & 0 & \bar{\eta} & 0 & 0 & 0
\end{array}\right] \begin{bmatrix} \beta_1 \\ \beta_2 \\ \vdots \\ \beta_{18} \end{bmatrix}. \tag{4.51}$$

Again, the analytic modes contain the patch test (β_1 to β_6) and bending modes (β_7 to β_{12}). Additionally, the three remaining linear modes (β_{13} to β_{15}) concern the shear components and cover states of pure torsion.[11] The final three modes are bilinear and were first proposed by Pian and Chen [111]. They are required to suppress zero energy modes in assumed stress elements. These modes fulfill equilibrium (4.2) with $\mathbf{b}^* = \mathbf{0}$ but do *not* lead to a compatible strain field in case of isotropy.[12] This is why they are labeled incompatible stress modes.

It would of course be better to find suitable compatible stress modes. Unfortunately, these do, to the best knowledge of the authors, not exist. Compatible modes would need either full quadratic fields or a priori knowledge of the material model. Both of which are not satisfactory which is why (4.51) is employed in most works (e.g., [38, 81, 112]). Other approaches [140, 145] perform either worse or at best similar compared to (4.51).

4.4.4.2. Trial functions for displacement and incompatible strain

The 3D metric shape functions M_i^e, $i = 1, \dots, 8$ for an eight node hexahedral element are defined analogously to (4.40) and contain the monomials $1, \bar{\xi}, \bar{\eta}, \bar{\zeta}, \bar{\eta}\bar{\zeta}, \bar{\zeta}\bar{\xi}, \bar{\xi}\bar{\eta}, \bar{\xi}\bar{\eta}\bar{\zeta}$. Again we obtain complete quadratic fields in skew coordinates by complementing the metric shape functions with quadratic incompatible modes $\tilde{M}_j^e$, $j = 1, 2, 3$. They are computed analogously to (4.45) using $\tilde{m}_1^e = \bar{\xi}^2$, $\tilde{m}_2^e = \bar{\eta}^2$ and $\tilde{m}_3^e = \bar{\zeta}^2$. As consequence of the complete quadratic displacement ansatz all linear stress modes (4.51) up to β_{15} can be exactly represented regardless of the material model.

Approaching the remaining bilinear stress modes β_{16} to β_{18} in the same way would require nine additional cubic incompatible displacement modes which are by far too many for an efficient element. Furthermore, the bilinear stress modes are incompatible and of subordinate importance due to their higher order nature. We therefore propose to follow the approach of Simo et al. [132] instead and focus on volumetric locking.

Andelfinger et al. [4] first observed that using only the three quadratic incompatible modes described above is not enough to eliminate volumetric locking. Following Simo et al. [132] this can be explained by looking at the trace of the strain tensor, which assumes the form

$$\begin{aligned} \operatorname{tr}(\boldsymbol{\varepsilon}^{\mathrm{h},e}) &= \operatorname{tr}(\nabla_{\mathbf{x}}^{s}\mathbf{u}^{\mathrm{h},e} + \tilde{\boldsymbol{\varepsilon}}^{\mathrm{h},e}) \\ &= C_0 + C_1\bar{\xi} + C_2\bar{\eta} + C_3\bar{\zeta} + C_4\bar{\eta}\bar{\zeta} + C_5\bar{\zeta}\bar{\xi} + C_6\bar{\xi}\bar{\zeta} \quad + \quad \tilde{C}_1\bar{\xi} + \tilde{C}_2\bar{\eta} + \tilde{C}_3\bar{\zeta}, \end{aligned} \tag{4.52}$$

where C_i and $\tilde{C}_i$ are constants depending on the displacement and enhanced degrees of freedom, respectively. The last equation immediately reveals that the incompressibility

[11] See, e.g., the work of [62] for analytic solutions associated with the linear modes.

[12] It is not possible to find a displacement field that leads to $\boldsymbol{\sigma}^* = \beta_{16}\, yz(\mathbf{e}_x \otimes \mathbf{e}_x)$. The compatibility condition $\varepsilon^*_{11,23} = \varepsilon^*_{13,21} + \varepsilon^*_{12,31} - \varepsilon^*_{23,11}$ is violated for the simple isotropic material which ultimately prohibits finding a suitable $\mathbf{u}^*$.

condition $\text{tr}(\boldsymbol{\varepsilon}^{\text{h},e}) = 0$ can only be met if $C_4 = C_5 = C_6 = 0$. This restriction leads to a mild form of locking which can be alleviated using three additional enhanced modes

$$\tilde{\boldsymbol{\varepsilon}}^{\text{h},e} = \mathbf{J}_0^{-\text{T}}(\alpha_{10}\bar{\xi}\bar{\eta}\mathbf{I} + \alpha_{11}\bar{\eta}\bar{\zeta}\mathbf{I} + \alpha_{12}\bar{\zeta}\bar{\xi}\mathbf{I})\mathbf{J}_0^{-1} \tag{4.53}$$

in the spirit of Simo et al. [132]. It is straightforward to verify this by observing that (4.52) now includes the additional terms $\tilde{C}_4\bar{\eta}\bar{\zeta} + \tilde{C}_5\bar{\zeta}\bar{\xi} + \tilde{C}_6\bar{\xi}\bar{\zeta}$ which relaxes the incompressibility conditions $C_m = 0$ to $C_m = -\tilde{C}_m$, $m = 1, 2, 3$. In contrast to Simo et al. [132] we use skew coordinates in (4.53) and do not apply the scaling by determinants. With the three additional modes (4.53) there are now in total twelve enhanced modes.

All in all, the three quadratic incompatible Wilson-modes formulated in terms of skew coordinates together with (4.53) enable exact representation of arbitrary linear stress fields and volumetric locking-free elements.

Remark 4.6. *Other approaches for volumetric locking-free EAS elements require either more internal degrees of freedom [1, 3, 55, 63] or are only slight modifications of* (4.53) *[67].*

4.4.4.3. Test function for the incompatible strain

The test functions for the incompatible strain are chosen in exactly the same way as in the work of Simo et al. [132], which are given by

$$\bar{\mathbf{e}}_v := \left[\begin{array}{ccc|cccccc|ccc} \xi & 0 & 0 & 0 & 0 & 0 & 0 & 0 & 0 & \xi\eta & \eta\zeta & \zeta\xi \\ 0 & \eta & 0 & 0 & 0 & 0 & 0 & 0 & 0 & \xi\eta & \eta\zeta & \zeta\xi \\ 0 & 0 & \zeta & 0 & 0 & 0 & 0 & 0 & 0 & \xi\eta & \eta\zeta & \zeta\xi \\ 0 & 0 & 0 & \xi & \eta & 0 & 0 & 0 & 0 & 0 & 0 & 0 \\ 0 & 0 & 0 & 0 & 0 & \eta & \zeta & 0 & 0 & 0 & 0 & 0 \\ 0 & 0 & 0 & 0 & 0 & 0 & 0 & \xi & \zeta & 0 & 0 & 0 \end{array}\right] \begin{bmatrix} \alpha_1 \\ \alpha_2 \\ \vdots \\ \alpha_{12} \end{bmatrix}, \tag{4.54}$$

along with the transformation formula in (4.47). Unfortunately however, (4.48) is not a priori fulfilled for the bilinear modes in (4.51) and (4.54). It is thus necessary to find a proper L_2-orthogonal set of test functions for the strain modes. This can be achieved by starting with an arbitrary $\bar{\mathbf{e}}_v$ and a two step *Gram-Schmidt* orthogonalization procedure. The stress modes are orthogonalized to each other in a first step before these modified stress modes are used to create an orthogonal enhanced field. It is crucial to conduct these operations using (4.48) instead of (4.29). This is because the latter leads to coupling between the different stress components which in turn induces frame dependence to the orthogonalized stress. The final orthogonalization in vector-matrix notation is described by

$$\bar{\mathbf{s}}^*_{v,i} = \bar{\boldsymbol{\sigma}}^*_{v,i} - \sum_{k=1}^{i-1} \frac{\int_{\hat{\Omega}} \bar{\boldsymbol{\sigma}}^*_{v,i} \cdot \bar{\mathbf{s}}^*_{v,k}\, j^{\text{h},e}\, \text{d}\hat{\Omega}}{\int_{\hat{\Omega}} \bar{\mathbf{s}}^*_{v,k} \cdot \bar{\mathbf{s}}^*_{v,k}\, j^{\text{h},e}\, \text{d}\hat{\Omega}} \bar{\mathbf{s}}^*_{v,k}, \qquad \bar{\boldsymbol{\epsilon}}^{\text{h},e}_{v,j} = \bar{\mathbf{e}}_{v,j} - \sum_{k=1}^{n_\sigma} \frac{\int_{\hat{\Omega}} \bar{\mathbf{e}}_{v,j} \cdot \bar{\mathbf{s}}^*_{v,k}\, \text{d}\hat{\Omega}}{\int_{\hat{\Omega}} \bar{\mathbf{s}}^*_{v,k} \cdot \bar{\mathbf{s}}^*_{v,k}\, j^{\text{h},e}\, \text{d}\hat{\Omega}} \bar{\mathbf{s}}^*_{v,k}, \tag{4.55}$$

where $\bar{\sigma}_{v,i}$ and $\bar{\mathbf{e}}_{v,j}$ denote columns of the ansatz matrices of $\bar{\sigma}_v^*$ and $\bar{\mathbf{e}}_v$, respectively. Furthermore, n_σ and n_ϵ are the number of stress and incompatible strain modes and $k = 1, \ldots, n_\sigma$ and $j = 1, \ldots, n_\epsilon$. The final ansatz for the test functions of the incompatible strain is given by

$$\tilde{\epsilon}^{\mathrm{h},e} = \frac{1}{j^{\mathrm{h},e}} \mathbf{J}_0^{-\mathrm{T}} \bar{\epsilon}^{\mathrm{h},e} \mathbf{J}_0^{-1} \tag{4.56}$$

and fulfills Condition C3 by construction.

4.4.4.4. Additional features

It is straightforward to show that the 3D element presented reduces to the linear version of the EAS element proposed by Simo et al. [132] in case of regular meshes.

Furthermore, as in Section 4.4.3.5, all integrands in (4.24) and (4.55) are of pure polynomial nature. However, standard $2 \times 2 \times 2$ Gauss quadrature is not sufficient for exact integration of the higher order expressions in (4.51), (4.53) and (4.54) in case of distorted elements. Fortunately though, all lower order modes are integrated exactly by standard $2 \times 2 \times 2$ Gauss quadrature and the under-integration of the higher order modes does not seem to affect the results as our numerical experiments show. It is furthermore interesting that increasing the order of quadrature does not lead to over-integration and too stiff response which is in contrast to standard finite elements.

Remark 4.7. *In case of generally distorted elements with non-constant Jacobians* $4 \times 4 \times 4$ *and* $3 \times 3 \times 3$ *Gauss quadrature is required to exactly integrate the expressions of the orthogonalization procedure* (4.55) *and the weak form* (4.24)*, respectively. Using different order of integration for the weak form and the orthogonalization procedure is not advisable as it leads to numeric violations of* (4.29)*. The required order of Gauss quadrature can be reduced to* $3 \times 3 \times 3$ *in the orthogonalization procedure* (4.55) *if one replaces the skew coordinates in the bilinear modes in* (4.51) *with parametric ones, which does not seem to influence the numeric results.*
However, we stress that despite not giving exact results, standard $2 \times 2 \times 2$ *Gauss points is sufficient for convergence and we thus employ it if not stated otherwise.*

4.5. Numerical investigations

This final section covers several numerical benchmarks comparing the novel finite element formulation to established ones. Special emphasize is put on mesh distortion sensitivity. To keep the descriptions of setup and results as concise as possible, we drop the 2D case in favor of the more complex 3D simulations. However, if not mentioned otherwise, results can qualitatively be transferred to 2D.

In the sequel we denote the present element **Q1U/E4** and **H1U/E12** in 2D and 3D, respectively. Therein, **U** denotes the unsymmetry of the stiffness matrix and **E***x* accounts for the number of enhanced modes. The standard elements we use for comparison are

- **H1** standard isoparametric displacement-based element with Lagrangian shape functions.
- **H1/E9** EAS element based on the nine Wilson-modes. The 2D version of this element, **Q1/E4**, was proposed in the seminal work of Simo and Rifai [134]. A 3D extension can, e.g., be found in the work of Andelfinger and Ramm [3].
- **HA1/E12** linearized form of the EAS element proposed by Pfefferkorn and Betsch [104] which is closely related to the improved EAS element by Simo et al. [132]. It uses three additional volumetric enhanced modes, a special nine point quadrature rule and a modified evaluation of the compatible displacement gradient.
- **H1/S18** assumed stress element with 18 stress modes as proposed by Pian and Tong [112].
- **H1/P0** the three field mixed pressure element with constant pressure and dilatation approximation as proposed by Simo et al. [135].
- **H1U/IM-S** the 3D extension of the unsymmetric incompatible mode element by Huang et al. [59]. We use a serendipity approach instead of the proposed Lagrange ansatz and perform the orthogonality procedure (4.55) to fulfill Condition C3, which has not been considered in Huang et al. [59].

Other low-order unsymmetric finite element formulations as the ones proposed in Cen et al. [32], Xie et al. [158] and Huang et al. [59] (only 2D) have also been tested. Their accuracy is very similar to the present element in most tests. However, they suffer from other severe drawbacks as outlined in the introduction and are therefore not included in this section. The only exemption is H1U/IM-S (see above) in the Cook's-membrane example in Section 4.5.5.

4.5.1. Patch test

The first benchmark is the classical patch test which is performed here as described in Pfefferkorn and Betsch [104, 105]. A block $\mathcal{B}_0 = [0,1]^3$ with Young's modulus $E = 10^4$ and Poisson's ratio $\nu = 0.3$ is subjected to boundary conditions $u_i(x_i = 0, x_j, x_k) = 0$, $i, j, k \in \{1, 2, 3\}$, $i \neq j \neq k$ and loaded via displacement $u_3(x_3 = 1, x_1, x_2) = u$. A regular mesh with $3 \times 3 \times 3$ elements as well as a distorted mesh with seven elements according to MacNeal and Harder [86] is used to discretize the block. Figure 4.5 shows the deformed configuration and stress distribution for both types of meshes.

By considering the stress distribution we can establish if the elements pass the patch test, that is, if the linear displacement states lead to the correct constant stress state. The nodal stresses of the novel H1U/E12 at a displacement state of $u = 0.6$ average to $\sigma = 6000$ for

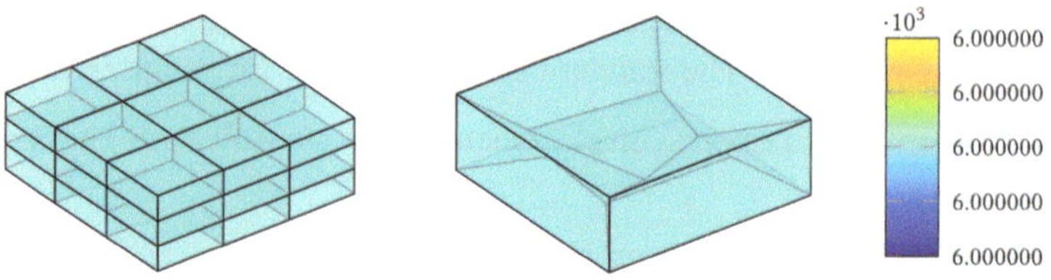

Figure 4.5.: Patch test. Deformed configuration ($u = 0.6$) with von Mises stress distribution for the regular (left) and distorted (right) mesh.

both meshes, which is the analytic solution. Moreover, the standard deviation of below 10^{-10} establishes the constant stress state. Thus, the novel element passes the patch test, which is not surprising since H1U/E12 was designed to pass the patch test by choice of the analytic stress modes (see Section 4.4.2).

Remark 4.8. *Other quadrature rules than the standard* $2 \times 2 \times 2$ *rule might induce failure of the patch test. Using, e.g.,* $2 \times 2 \times 2$ *and* $3 \times 3 \times 3$ *Gauss points for integration of the residual* (4.24) *and orthogonalization procedure* (4.55)*, respectively, leads to a violation of the patch test. An equal number of Gauss points for both the residual and the orthogonalization (e.g., the* $2 \times 2 \times 2$ *for H1U/E12) ensures satisfaction of the patch test. Alternatively, an exact integration, which requires at least* $3 \times 3 \times 3$ *and* $4 \times 4 \times 4$ *Gauss points, respectively, also ensures that the patch test is passed.*

4.5.2. Isotropy and frame-indifference

With this simple one element test we check whether the novel element formulation fulfills two basic properties. First, we test for frame-indifference with respect to rotation of the coordinate system and second for isotropy with respect to choice of the local element axes. This test was proposed in [140, 158] and is slightly modified here.[13]

We consider a single finite element of general shape whose geometry in the local $\bar{x}\bar{y}\bar{z}$ coordinate frame is described in Figure 4.6. Dirichlet boundary conditions $\mathbf{u} = \mathbf{0}$ apply on nodes A, D, E, H and two nodal forces act on nodes C and G as shown in Figure 4.6. The material properties are $E = 1500$ and $\nu = 0.25$. To test for frame-indifference we consider the case of no rotation, that is, the local $\bar{x}\bar{y}\bar{z}$-frame coincides with the global xyz-frame, and a case where the element is rotated by 65°, 15° and 25° around the x, y and z axis, respectively. Furthermore, we use different element node-numberings (see Figure 4.6) to change the local element axes with respect to the coordinate frame $\bar{x}\bar{y}\bar{z}$. If this does not influence the result the element is isotropic.

[13] In particular, we use a different geometry to ensure generality of the test. The nodes chosen in [140, 158] are symmetric to the $\bar{x}\bar{z}$-plane which should be avoided.

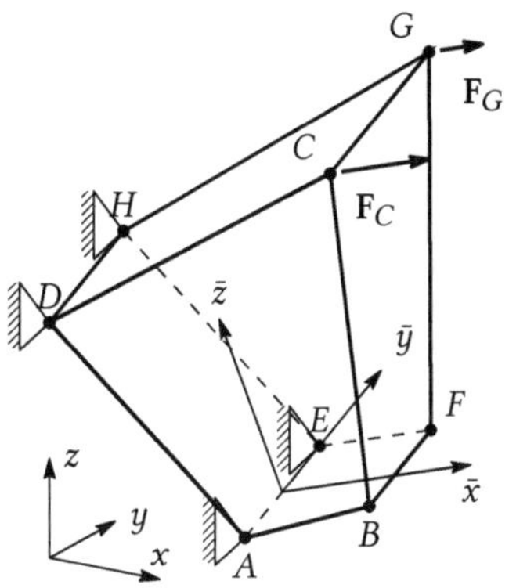

Nodes and forces in $\bar{x}\bar{y}\bar{z}$ coordinates:

$A = (0, -0.75, 0)$	$E = (0, 0.75, 0)$
$B = (1, -0.50, 0)$	$F = (1, 0.75, 0)$
$C = (2, -1, 3)$	$G = (2, 1, 3)$
$D = (-1, -0.75, 2)$	$H = (-1, 0.75, 2)$
$\mathbf{F}_C = [200, 0, 0]^{\mathrm{T}}$	$\mathbf{F}_G = [100, 0, 0]^{\mathrm{T}}$

Element node-numberings:

$\mathrm{EDOF}_1 = [A, B, C, D, E, F, G, H]$

$\mathrm{EDOF}_2 = [F, B, C, G, E, A, D, H]$

$\mathrm{EDOF}_3 = [A, D, C, B, E, H, G, F]$

Figure 4.6.: Isotropy and frame-indifference. Setup of the test in 3D.

All standard elements and the novel element H1U/E12 pass this test and are indifferent to rotation as well as node-numbering. Specifically, the novel element exhibits a displacement of $u = 1.3696$ for all six combinations of rotation and node-numbering. The standard deviation of the six displacements is $5.4 \cdot 10^{-16}$.

Remark 4.9. *Isotropy and frame-indifference of H1U/E12 are a direct consequence of using skew coordinates as described in Section 4.4.1 which are extensively used for both trial fields $\mathbf{u}^{h,e}$ and $\tilde{\varepsilon}^{h,e}$. If we used the centroid coordinates $\bar{\mathbf{x}} = \mathbf{x}^{h,e} - \mathbf{x}_0$ instead of the skew coordinates* (4.32)*, the element would loose its frame-indifference. In our numerical test this is established by the increased standard deviation of $1.7 \cdot 10^{-3}$. This frame-dependence of the centroid coordinates has first been cured by a rotation [100]. Unfortunately, that approach induces anisotropy (see Xie et al. [158]). So far, the only proper choice is using the skew coordinates which simultaneously scale and rotate the centroid coordinates, see* (4.32).

4.5.3. Eigenvalue analysis

Next we perform the eigenvalue analysis of the stiffness matrix as proposed by Simo et al. [132]. A single finite element with regular or distorted shape as shown in Figure 4.7 is considered for this test. The shear and bulk modulus are $\mu = 1$ and $\kappa = 10^9$ which corresponds to nearly incompressible material behavior and causes volumetric locking for standard elements.

After static condensation, the stiffness matrix of the single finite element $\mathbf{K}^e$ is always of the size 24×24. Thus, $\mathbf{K}^e$ has 24 eigenvalues which can be grouped into four sets. First, zero eigenvalues $\lambda_i = 0$ which correspond to rigid body modes. Second, locking modes with $\lambda_i \longrightarrow +\infty$. In the numeric examples infinity is not reached and therefore modes with "high" eigenvalues are considered as locking modes. The remaining eigenpairs are normal modes if their eigenvalues $\lambda_i > 0$ are positive and instable if $\lambda_i < 0$. The desired result is that the element exhibits the correct number of six rigid body modes,

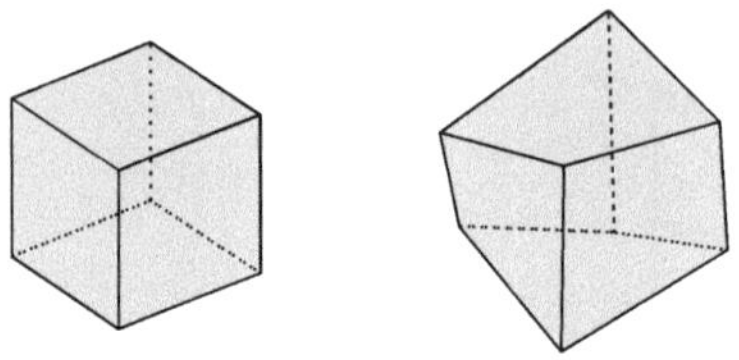

i	1	2	3	4	5	6	7	8
x	−1.249	1.317	−0.821	0.961	−0.941	0.937	−1.356	1.148
y	−1.195	−0.923	1.221	0.712	−0.846	−1.278	0.869	1.054
z	−1.114	−0.865	−0.789	−1.261	0.963	1.056	0.745	1.312

Figure 4.7.: Eigenvalue analysis. Regular $\Omega^{\text{reg}} = [-1, 1]^3$ (top) and distorted Ω^{dist}(bottom) cube for the eigenvalue analysis. Table with positions of nodes $\mathbf{x}_i$ of the distorted element Ω^{dist} [104].

has no instable modes $\lambda_i < 0$ and exactly one locking mode which corresponds to pure dilatational deformation and accounts for the incompressibility. The other modes should be normal modes with $\lambda_i > 0$.

The number of modes in each category for the elements tested are shown in Table 4.1. Apart from elements H1 and H1/E9 all elements are free from locking in this test and exhibit the desired behavior. Failure to reproduce the correct number of modes immediately establishes that the element suffers from locking. For example, the four and five locking modes of H1/E9 in the regular and distorted case result in a mild form of locking which influences, e.g., limit loads in elasto-plastic simulations as first reported by Andelfinger et al. [4]. Unfortunately, passing this test is not sufficient to prove that an element is locking free. It is for instance well-known that H1/P0 suffers from shear locking yet it passes the test. It is much more difficult to detect shear locking using this simple test. Even if the block is scaled in one direction to mimic a shell-like geometry there are no additional locking modes for H1/P0. The only difference compared to shear locking-free elements is that a few eigenvalues are slightly higher but not clearly distinguishable.

The novel element H1U/E12 passes this test and exhibits the correct number of modes in each category. It is also shear locking-free which we show with subsequent examples. However, due to the unsymmetric stiffness matrix it is not guaranteed that all eigenvalues are real-valued. Indeed there is one pair in the distorted case with $\lambda_i = 2.177 \pm 0.02374i$ where the imaginary part is not negligibly small. So far, there seems to be no consequence of this in all the simulations we conducted. Interestingly, no imaginary eigenvalues occur in the 2D case even though the stiffness matrix is unsymmetric as well.

Table 4.1.: Number of eigenvalues of the stiffness matrix in each category.

	regular element			distorted element		
element type	rigid body[a]	normal[b]	locking[c]	rigid body[a]	normal[b]	locking[c]
H1	6	11	7	6	10	8
H1/E9	6	14	4	6	13	5
HA1/E12T	6	17	1	6	17	1
H1/P0	6	17	1	6	17	1
H1/S18	6	17	1	6	17	1
H1U/E12	6	17	1	6	17	1

[a] rigid body modes with $|\lambda_i| \leq 10^{-6}$
[b] normal modes with $0.1 \leq \lambda_i \leq 10$
[c] locking modes with $\lambda_i \geq 1000$ (most satisfy $\lambda_i \geq 10^8$)

4.5.4. Mesh distortion

To examine the influence of mesh distortion we consider a 3D version [104] of the widely used mesh distortion test (e.g., [3, 32, 59, 134]). The cantilevered beam-like structure is subjected to pure bending as shown in Figure 4.8. The specimen of length $l = 10$, height $h = 2$ and width $b = 1$ is meshed with only two elements and distortion is applied via parameter s. At the clamped end of the beam boundary conditions $u(x = 0, y, z) = 0$, $v(x=0, y=0, z) = 0$ and $w(x=0, y, z=0) = 0$ apply. The free end is subjected to a bending moment $M = 20$ which is applied via traction boundary condition $\sigma(z) = 30 \cdot (1 - z)$. We use material parameters $E = 1500$ and $\nu = 0.25$. An analytic solution exists for this problem and can, e.g., be found in the work of Jabareen and Rubin [62]. The exact value for displacement δ in z-direction at point $(10, 0.5, 2)$ is $\delta = 1$.

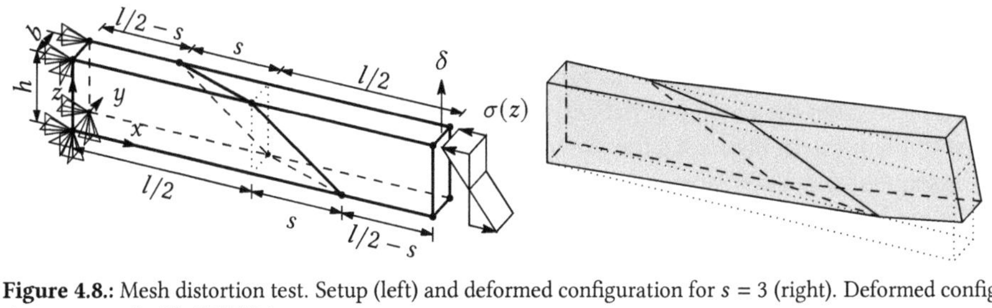

Figure 4.8.: Mesh distortion test. Setup (left) and deformed configuration for $s = 3$ (right). Deformed configuration computed with H1U/E12.

Results of the mesh distortion test are shown in Figure 4.9 where we compare the numerically computed displacement δ for increasing skew s. We limit the degree of skew to $s = 4.9$ instead of the usual $s = 5$ since the matrix needed to compute the metric shape functions in (4.40) becomes singular for coincident nodes.

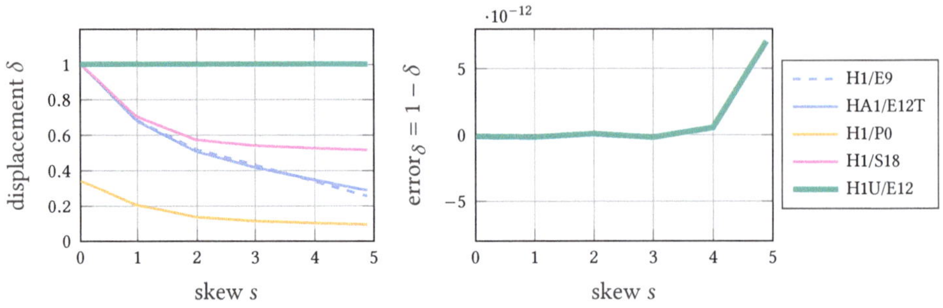

Figure 4.9.: Mesh distortion test. Displacement δ for different degrees of skew s (left) and error of displacement compared to analytic solution (right, only element H1U/E12).

Apart from element H1/P0, which suffers from shear locking, all elements are exact for a regular mesh with $s = 0$. However, for distorted meshes the computed displacement of the standard mixed finite elements drops to less than 60% of the required $\delta = 1$ for distortions $s \geq 2$. Of the standard finite elements the best results can be obtained with the assumed stress element H1/S18. There exist modified symmetric assumed stress elements that exhibit even better accuracy and show exact displacement for some nodes. We mention the elements proposed by Wu and Chueng [157] and Sze [139]. However, we stress that since the displacement is not exact at all nodes (see Table VI in the work [139]), these elements do not breakthrough MacNeal's theorem.
In contrast to the standard symmetric elements the novel H1U/E12 is exact *regardless* of the mesh distortion parameter s. The second plot in Figure 4.9 shows that there is only a small numerical round-off error compared to the analytic solution. This error slightly increases for higher distortions since the matrix needed to compute the metric shape functions is less well conditioned for highly distorted meshes. Moreover, the newly proposed element does not only reproduce the correct displacement at every node but is even pointwise exact. This holds for the displacement, the stress *and* the strain.

As pointed out in Section 4.2.2.2, it is not possible to obtain exact solutions to quadratic problems with low-order elements for all geometries. We can observe this by scaling the distorted mesh from the patch test in Section 4.5.1 for this test (see Figure 4.10, mesh 1). In contrast to the mesh in Figure 4.8 the skew axes of the elements do not align for this mesh. As a consequence, the displacement $\delta = 1.004$ obtained with H1U/E12 differs marginally from the analytic value $\delta = 1$. However, it is still a much better approximation than any of the standard elements can provide. Of those H1/S18 shows the best accuracy with $\delta = 0.146$, which is more than 85% off in contrast to the 0.4% of H1U/E12 (see Figure 4.11). Furthermore, H1U/E12 exhibits the analytic results in case of the slightly regularized mesh 2 which is shown in Figure 4.10 on the right. For that mesh the skew axes in longitudinal direction of the beam align which implies that pure bending around the y-axis is again part of the ansatz space.

All in all, the novel Petrov-Galerkin EAS element H1U/E12 was designed to be accurate for bending problems and therefore it performs extraordinarily well in this test.

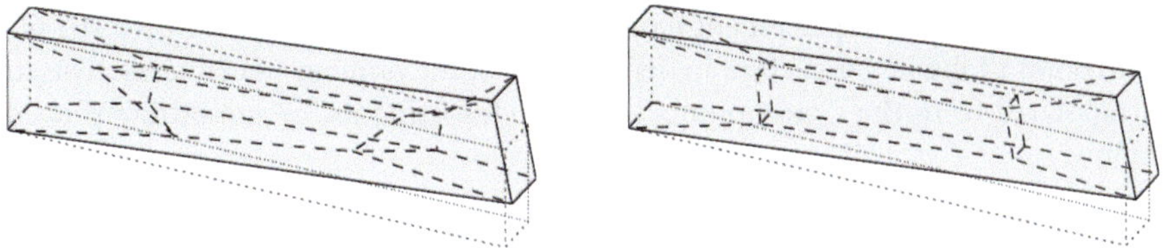

Figure 4.10.: Mesh distortion test. Special meshes and corresponding deformed configurations computed with H1U/E12. Scaled distorted patch test mesh 1 (left) and regularized version of the distorted patch test mesh 2 (right).

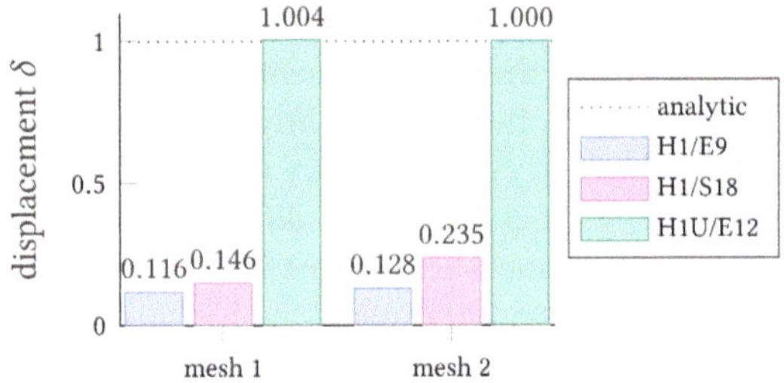

Figure 4.11.: Mesh distortion test. Displacement δ for special meshes 1 and 2.

4.5.5. Cook's membrane

The final numerical example in the present work is the Cook's membrane (e.g., [104, 105, 127]) which we use to investigate coarse mesh accuracy and convergence of results with mesh refinement. The tapered trapezoidal specimen shown in Figure 4.12 is clamped ($\mathbf{u}(x=0,y,z) = \mathbf{0}$) on the left and subjected to equally distributed shear stress $\tau = 100$ in y-direction on the right face. We assume near incompressibility and set the material parameters to $E = 2261$ and $\nu = 0.4955$. There are always two elements in direction of the thickness while we gradually refine the mesh in the other two directions using $n_{\text{el}} = \{2, 4, 8, 16\}$ elements per side.

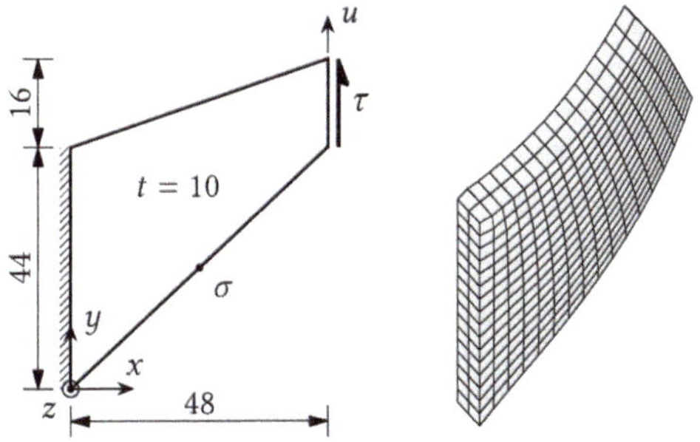

Figure 4.12.: Cook's membrane. Geometry, boundary conditions (left) and deformed configuration computed using H1U/E12 and a 16 × 16 × 2 mesh (right).

Figure 4.13 shows the displacement u of the top right corner and von Mises stress σ at the mid point of the lower surface in dependence of the number of elements. The stress is extrapolated from the Gauss points to the nodes using an L_2-smoothing procedure given by

$$\sum_{e=1}^{n_{\mathrm{el}}} \tau_i \int_{\Omega^e} N_i N_j \,\mathrm{d}V \sigma_j = \sum_{e=1}^{n_{\mathrm{el}}} \tau_i \int_{\Omega^e} N_i \hat{\sigma} \,\mathrm{d}V, \qquad \forall\, \tau_i \in \mathbb{R}. \tag{4.57}$$

Therein, N_i are the standard parametric shape functions, σ_j the stress at the nodes and $\hat{\sigma}$ the constitutive stress[14] within the element. We use the same Gauss points for the stress extrapolation as for the integration of element stiffness matrices and residuals.
The plots in Figure 4.13 show that the novel element exhibits the most accurate displacement and stress for all mesh sizes. It is even slightly better than the other unsymmetric element H1U/IM-S. Especially the accuracy for very coarse meshes is extremely high compared to the other elements. For finer meshes the difference becomes smaller because the deformation state within single elements converges towards the patch test state for which all elements are exact. Furthermore, the principle of limitation discussed in Remark 4.2 prohibits higher order convergence beyond $\mathcal{O}(h^2)$ in the limit of fine meshes.
Nevertheless, the present example shows that the novel element can also improve results if the sought displacement state is not part of the ansatz space. However, the improvements are in general not as impressive as in bending dominated examples. After all, it is a *low-order* numerical method with its limitations.

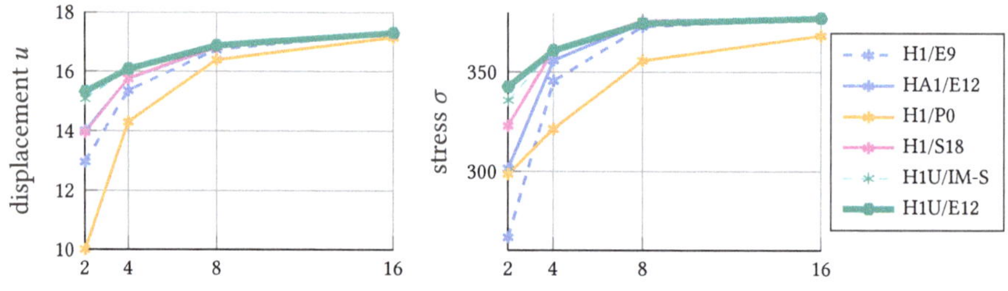

Figure 4.13.: Cook's membrane. Convergence of displacement u (left) and stress σ (right) with mesh refinement.

Another aspect we examine with the Cook's membrane benchmark is the computational effort needed for the newly proposed H1U/E12 in comparison with the symmetric element HA1/E12. To that end we focus on the time taken to solve the global linear equation system since it usually requires most of the computing time of a finite element program [154]. We employ MATLAB's [90] "\"-operator, which automatically selects a suitable efficient and highly optimized algorithm. In the present case it uses a *sparse* Cholesky and LU-factorization for the symmetric and unsymmetric stiffness matrix, respectively. Other parts

[14] For H1/S18 we use the independent stress field.

of the finite element program are of subordinate importance and are either equivalent for both methods (assembly, pre- and post-processing) or hardly comparable due to non-optimized code (element-routine, computation of shape functions). These parts are therefore not considered here. Figure 4.14 shows the median relative computation time of H1U/E12 compared to HA1/E12 of 50 runs for different mesh sizes with up to 114075 degrees of freedom. The run times are highly volatile in particular for coarse meshes. There are many outliers which is why we considered the median instead of the mean value. Nevertheless, Figure 4.14 gives an idea of the additional computational effort which seems to be relatively independent of the mesh size and is about four times the effort of a symmetric method for the same mesh size. However, as Figure 4.13 and the other examples of the present work suggest, less elements might be required with the novel approach.

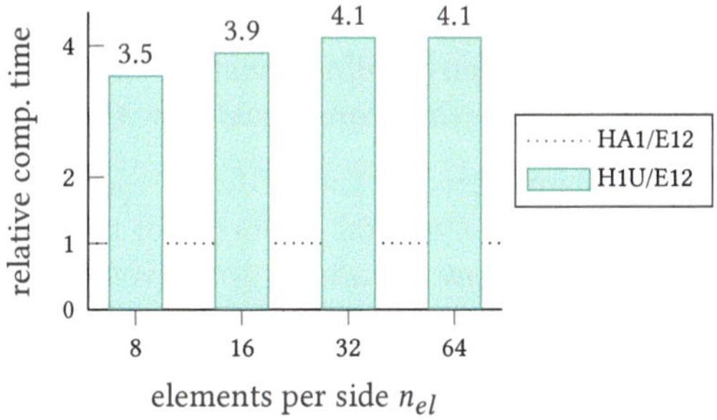

Figure 4.14.: Cook's membrane. Comparison of the computing time required for the solution of the linear system of equations.

4.6. Conclusion

We designed a novel high performance *Petrov-Galerkin* mixed finite element based on the EAS framework. The rather uncommon Petrov-Galerkin approach is motivated by MacNeal's theorem [84], which is revisited in the present work and in particular states that the optimal element performance can only be obtained with an unsymmetric stiffness matrix.

Within the newly-proposed Petrov-Galerkin EAS framework, we devised three design conditions which ensure that the finite element is capable of exactly reproducing nodal displacements corresponding to specific deformation modes. The proposed element is designed to fulfill these conditions for rigid body, patch test and bending modes, which are of the utmost importance in many engineering applications. For the construction of the ansatz spaces, skew coordinates play a crucial role to obtain the novel frame-indifferent element with exceptional coarse mesh accuracy and low sensitivity to mesh distortion. In case of the patch test and bending problems, for which the element is specifically crafted, the element's response is even pointwise exact not only for the displacement but also

for the stress and strain. Furthermore, in case of regular meshes it is equivalent to the standard EAS approach.
Previously developed low-order unsymmetric elements suffer from major drawbacks that could be overcome with the present Petrov-Galerkin EAS-framework. In particular, the new approach does neither involve ansatz functions dependent on material parameters [32, 158] nor does it require many internal degrees of freedom and/or higher order numerical integration [59]. Its extension to 3D is also straightforward in contrast to many existing elements.
In practical simulations the key feature of the novel EAS-framework is probably its reduced mesh distortion sensitivity. This might help to reduce the effort to generate finite element meshes since the high quality demands on mesh topology can be reduced.
A drawback of unsymmetric elements in general is the increased numerical effort compared to standard methods. This concerns in particular the computation of the ansatz functions and the factorization of the unsymmetric stiffness matrix. However, this additional effort is in the opinion of the authors more than compensated by the increased accuracy.

Future work should first focus on extending the method presented to nonlinear problems. Even though it is not possible to find general solutions for the large variety of material models, our preliminary results show significant improvements, especially regarding sensitivity to mesh distortion. Other interesting lines of research are the extensions to shell problems as well as dynamics.

Appendix to Chapter 4

4.A. Generalizations of MacNeal's theorem

4.A.1. MacNeal's theorem with $\mathbf{b}^* \neq \mathbf{0}$

The first simplification made in Section 4.2 was considering only modes without body force (4.5). However, higher order modes can in general not be represented without appropriate $\mathbf{b}^*$ computed using (4.2). These non-zero body forces $\mathbf{b}^* \neq \mathbf{0}$ have to be considered in (4.6). Fortunately, it is still possible to write the external part of the weak form as

$$G_{\text{ext}}^{*,e} = \sum_{m,n=1}^{M} \beta_m F_{mn}^* \alpha_n, \tag{4.58}$$

where F^*_{mn} now includes contributions form $\mathbf{t}^*$ and $\mathbf{b}^*$. Since there is no change of the internal part (4.8), all conclusions on the structure of F^*_{mn} remain the same. Thus it is symmetric and by proper choice of $\mathbf{u}^*_m$ diagonal. The external part of the weak form from the finite element solution (4.17) changes accordingly with the only necessary change being the addition of the body force integral. Furthermore, there is no change in the conclusions drawn for nodal equilibrium. This follows from the fact that the body force concerns only the internal of the element while nodal equilibrium is determined by balancing the forces on the surface.

All in all, it is straightforward to consider modes with $\mathbf{b}^* \neq \mathbf{0}$ by simply adding the missing integral in all external parts of the weak form.

4.A.2. Petrov-Galerkin approach

Choosing the Bubnov-Galerkin approach in Section 4.2.1.2 does not limit the generality of the present proof. Any proper set of M linear independent modes $\mathbf{v}^*_m$ for the test function yields in general an unsymmetric F^*_{mn} as (4.6) and (4.8) show. However, by use of the *singular value decomposition* and transformation of the weights α_n, β_m using the unitary matrices of said decomposition, the virtual work of the external forces can always be expressed in the form (4.6) with a diagonal matrix F^*_{mn}.

Thus, it is possible to get the same structure with the Petrov-Galerkin approach as with the Bubnov-Galerkin approach and all further conclusions apply.

4.A.3. Generalized displacement approximations

As mentioned in Section 4.2.2.1 the approximation (4.11) is not very general and especially excludes cases of high practical importance where M^e_i do not have the Kronecker-delta property and thus $\mathbf{u}^e_i$ are not the nodal displacements. Widely used examples of such non-interpolatory approaches are *hierarchical* higher order elements [162] and the *isogeometric* analysis [34]. Fortunately, it is still possible to find unique $\mathbf{u}^{*,e}_{im}$ such that $\mathbf{u}^e_i$ can be computed according to (4.12). Considering an even more general displacement ansatz with matrix-valued ansatz functions such that

$$\mathbf{u}^{\mathrm{h},e}(\mathbf{x}) = \sum_{i=1}^{N} \mathbf{M}^e_i(\mathbf{x})\mathbf{u}^e_i \tag{4.59}$$

still allows to compute $\mathbf{u}^e_i$ from equation system $\mathbf{u}^{\mathrm{h},e}(\mathbf{x}_j) = \mathbf{u}^*(\mathbf{x}_j)$, $j = 1, \dots, N$ if the fixed locations $\mathbf{x}_j$ are chosen appropriately (usually the nodes). Examples for matrix valued ansatz functions for the displacement can be found, e.g., for Petrov-Galerkin finite elements in [32, 76, 158, 161] or as special cases of incompatible mode elements in [22].

The general approximation (4.59) can also be applied to the test function. If this is done, the forces $\mathbf{P}^{\mathrm{h},e}_i$ defined in (4.17) should not be termed nodal force anymore since they do not directly correspond to a single node anymore due to the lost Kronecker-delta property. Accordingly, the expression *nodal equilibrium* should not be used. We use the expression generalized nodal force and equilibrium instead. Apart from this terminology, however, there are no changes.

In view of the restrictions due to equilibrium of generalized nodal forces, which apply accordingly, it is questionable if matrix valued functions are sensible for the test functions of the displacement. As shown in Section 4.2.2.3 nodal equilibrium severely restricts the choice of ansatz functions and one does not gain much flexibility with the matrix valued functions if standard assumption (4.21) is made. Furthermore, the matrix valued functions are restricted due to isotropy requirements and must be chosen such that changes of coordinate system do not alter the results. Ultimately, we conclude that using scalar functions for the test function of the displacement is most likely the only reasonable choice.

4.B. Skew coordinate frame

This Appendix describes how the direct link between the skew coordinates $\bar{\boldsymbol{\xi}}$ and the isoparametric coordinates $\boldsymbol{\xi}$ given in (4.35) can be established. We present the 3D case which can easily be reduced to the corresponding 2D formulation. By virtue of (4.32) and (4.30) we get in a first step

$$\bar{\boldsymbol{\xi}} = \mathbf{J}_0^{-1} \left(\sum_{i=1}^{8} N_i(\boldsymbol{\xi}) \mathbf{x}_i^e - N_i(\mathbf{0}) \mathbf{x}_i^e \right). \tag{4.60}$$

Now, we use an alternative representation of shape functions frequently used in the context of hourglass-stabilization (e.g., [104, 154]). The vector of all eight Lagrangian shape functions is given by

$$\mathrm{N} = \frac{1}{8} \mathbf{a}^0 + \frac{1}{8} \sum_{I=1}^{3} \mathbf{a}^I \xi_I + \frac{1}{8} \sum_{A=1}^{4} \mathbf{h}^A H_A(\boldsymbol{\xi}). \tag{4.61}$$

where

$$H_1 = \eta\zeta, \qquad H_2 = \xi\zeta, \qquad H_3 = \xi\eta, \qquad H_4 = \xi\eta\zeta, \tag{4.62}$$

$$\begin{aligned}
\mathbf{a}^0 &= \begin{bmatrix} +1 & +1 & +1 & +1 & +1 & +1 & +1 & +1 \end{bmatrix}^{\mathrm{T}}, & \mathbf{h}^1 &= \begin{bmatrix} +1 & +1 & -1 & -1 & -1 & -1 & +1 & +1 \end{bmatrix}^{\mathrm{T}}, \\
\mathbf{a}^1 &= \begin{bmatrix} -1 & +1 & +1 & -1 & -1 & +1 & +1 & -1 \end{bmatrix}^{\mathrm{T}}, & \mathbf{h}^2 &= \begin{bmatrix} +1 & -1 & -1 & +1 & -1 & +1 & +1 & -1 \end{bmatrix}^{\mathrm{T}}, \\
\mathbf{a}^2 &= \begin{bmatrix} -1 & -1 & +1 & +1 & -1 & -1 & +1 & +1 \end{bmatrix}^{\mathrm{T}}, & \mathbf{h}^3 &= \begin{bmatrix} +1 & -1 & +1 & -1 & +1 & -1 & +1 & -1 \end{bmatrix}^{\mathrm{T}}, \\
\mathbf{a}^3 &= \begin{bmatrix} -1 & -1 & -1 & -1 & +1 & +1 & +1 & +1 \end{bmatrix}^{\mathrm{T}}, & \mathbf{h}^4 &= \begin{bmatrix} -1 & +1 & -1 & +1 & +1 & -1 & +1 & -1 \end{bmatrix}^{\mathrm{T}}.
\end{aligned} \tag{4.63}$$

Inserting these relations into (4.60) together with $\tilde{\mathbf{x}}^e = \begin{bmatrix}\mathbf{x}_1^e & \cdots & \mathbf{x}_8^e\end{bmatrix}$ allows to recast (4.60) as

$$\begin{aligned}\bar{\xi} &= \mathbf{J}_0^{-1}\tilde{\mathbf{x}}^e\,[\mathbf{N}(\xi) - \mathbf{N}(\mathbf{0})] = \mathbf{J}_0^{-1}\tilde{\mathbf{x}}^e \left[\frac{1}{8}\mathbf{a}^0 + \frac{1}{8}\sum_{I=1}^{3}\mathbf{a}^I\xi_I + \frac{1}{8}\sum_{A=1}^{4}\mathbf{h}^A H_A(\xi) - \frac{1}{8}\mathbf{a}^0\right] \\ &= \mathbf{J}_0^{-1}\left[\left(\sum_{i=1}^{8}\mathbf{x}_i^e \otimes \nabla_\xi N_i\big|_{\xi=0}\right)\xi + \frac{1}{8}\sum_{A=1}^{4}\sum_{i=1}^{8}\mathbf{x}_i^e h_i^A H_A(\xi)\right],\end{aligned} \tag{4.64}$$

where use has been made of the fact that

$$\nabla_\xi N_i\big|_{\xi=0} = \frac{1}{8}\begin{bmatrix}\mathbf{a}_1^{\mathrm{T}}\\ \mathbf{a}_2^{\mathrm{T}}\\ \mathbf{a}_3^{\mathrm{T}}\end{bmatrix}. \tag{4.65}$$

The term in parentheses in the last expression of (4.64) is $\mathbf{J}_0$. The last step is to introduce auxiliary variable

$$\mathbf{c}_A := \frac{1}{8}\sum_{i=1}^{8}\mathbf{x}_i^e h_i^A, \qquad A = 1, 2, 3, 4, \tag{4.66}$$

which only depends on the nodal positions. With this information at hand we can now cast the skew coordinates in the form

$$\bar{\xi} = \xi + \mathbf{J}_0^{-1}\sum_{A=1}^{4}\mathbf{c}_A H_A(\xi), \tag{4.67}$$

which is the 3D equivalent of (4.35). The final step to get the 2D form given in (4.35) is to adopt the equation above for the 2D case. There is then only one hourglass vector $\mathbf{h}^1 = \begin{bmatrix}+1 & -1 & +1 & -1\end{bmatrix}^{\mathrm{T}}$ and function $H_1 = \xi\eta$. Furthermore, the factor $\frac{1}{8}$ in (4.66) has to be replaced by $\frac{1}{4}$.

5. Hourglassing- and locking-free mesh distortion insensitive Petrov-Galerkin EAS element for large deformation solid mechanics

This chapter reproduces:[*]
Pfefferkorn R and Betsch P. "Hourglassing- and Locking-Free Mesh Distortion Insensitive Petrov-Galerkin EAS Element for Large Deformation Solid Mechanics". In: *Int J Numer Meth Eng*. 124(6): 1307–1343, 2023. DOI: 10.1002/nme.7166

Abstract: We present a novel geometrically nonlinear EAS element with several desirable features. First, a Petrov-Galerkin ansatz significantly improves the element's performance in distorted meshes without loosing the simple strain-driven format. Second, the recently proposed mixed integration point strategy is employed to improve the element's robustness in the Newton-Raphson scheme. Finally and most importantly, we enhance the spatial displacement gradient instead of the usually modified deformation gradient. This allows to construct an element without the well-known spurious instabilities in compression and tension as shown numerically and supported by a corresponding hypothesis. All in all, this leads to a robust, stable, locking-free and mesh distortion insensitive finite element successfully applied in a wide range of examples.

Keywords: enhanced assumed strain (EAS), Petrov-Galerkin, unsymmetric FEM, hourglassing instabilities, mixed finite elements, mixed integration point

5.1. Introduction

Countless mixed finite elements have been developed throughout the last 50 years with the goal to cure *locking* and other defects of low-order finite elements. One of the probably most successful element groups are *enhanced assumed strain* (EAS) elements due to their simplicity and strain driven format which allows straightforward implementation of complex material models. The EAS framework has first been proposed in the early 1990s by

Simo and Rifai [134] and Simo and Armero [131] for linear and nonlinear problems, respectively, and is a mathematically sound successor of the earlier proposed incompatible mode models [141, 149]. Since then, the EAS method has been implemented into several commercial codes and has not only been successfully applied in solid mechanics (e.g., [36, 47, 66–70, 95, 104–106, 109, 131, 132, 138, 155]) but also to model shell structures, diffusion problems and even fracture problems (e.g., [3, 9, 17, 28, 71, 80]). Despite the tremendous effort put in development of EAS elements there are still some open issues. Pfefferkorn and Betsch [107] discuss three of which: robustness in the Newton-Raphson scheme, mesh distortion sensitivity and hourglassing instabilities. All three are addressed in this contribution.

The first issue concerns *robustness* in the Newton-Raphson scheme by which we herein denote two properties: maximum size of applicable load steps and number of Newton-Raphson iterations required for convergence [109]. In this regard assumed stress elements are by far superior to both displacement-based elements [45, 87] and EAS elements [109, 145]. To improve robustness of displacement-based elements Magisano et al. [88] recently proposed the *mixed integration point* (MIP) method in the context of shell problems which allows to substantially increase the robustness by a simple modification. It only affects the finite element tangent without changing the residual (and therefore the solution). This approach has recently been extended to EAS elements (see Pfefferkorn et al. [109]) and is also employed in the present contribution in order to improve robustness of the Petrov-Galerkin EAS approach at hand.

Regarding the second issue, *mesh distortion sensitivity*, there exists a fundamental publication by MacNeal [84] (see also the preliminary work [83]) which has in the opinion of the authors not gotten the attention it deserves. In that work MacNeal [84] *proves* (see also Pfefferkorn and Betsch [106]) that a finite element for linear elasticity with symmetric stiffness matrix *cannot* simultaneously satisfy the patch test and be exact for higher order modes for arbitrary meshes. This ultimately leads to mesh distortion sensitivity of elements with symmetric stiffness matrices. However, MacNeal's theorem leaves a "loop-hole". Elements with an unsymmetric stiffness matrix can simultaneously pass the patch test and be mesh distortion insensitive. Constructing such finite elements is the key idea of the *unsymmetric finite element method* first proposed by Rajendran and Liew [117] for higher order elements (see also [100, 116, 158]). That method is based on a *Petrov-Galerkin* approach instead of the usual *Bubnov-Galerkin* ansatz. Essentially, the trial functions of the displacement (actual field) are approximated with so-called *metric shape functions* [85] (constructed in the physical domain) while the test functions of the displacement (virtual field) are approximated as usual with parametric shape functions. This approach ultimately allows to greatly reduce the mesh distortion sensitivity.
A major drawback of all unsymmetric finite elements is of course the increased computational cost. This concerns especially the factorization of the non-symmetric stiffness matrix. The time required for that step is compared to standard elements in the work [106]. However, the additional effort is, in the opinion of the authors, more than compensated by the increased coarse mesh accuracy and the reduced mesh distortion sensitivity.
Unfortunately, it is not as straightforward to construct low-order unsymmetric finite elements because a careful choice of ansatz functions is required for optimal performance

[32, 106]. This lead previously to methods with complex ansatz functions that depend on the material model [32, 76–78, 158, 161] or that require higher order numerical integration [59]. Unsymmetric finite element approaches without these drawbacks have recently been published by Pfefferkorn and Betsch [106] in the form of a *Petrov-Galerkin EAS* (PG-EAS) framework and by Huang et al. [58] on the basis of an incompatible mode approach. Both elements are developed for linear elasticity and are much less distortion sensitive than any standard finite elements. Moreover, the approach by Pfefferkorn and Betsch [106] is completely locking-free, enables exact integration of the finite element integrals with standard Gauss quadrature and reduces to the standard EAS method in case of regular meshes. In this work we straightforwardly apply the unsymmetric PG-EAS approach by Pfefferkorn and Betsch [106] to nonlinear problems and show that it has similar beneficial effects. Mesh distortion sensitivity is greatly reduced even though it is not possible to get analytic results as in the linear elastic case. Furthermore, we propose a modification of the three volumetric modes required in 3D for a completely locking-free element [106]. This adjustment allows to skip the computationally expensive Gram-Schmidt orthogonalization required in aforementioned reference and is motivated by higher-order analytic solutions of linear elasticity similar to the ones given by Nadler and Rubin [96].

The final and probably most important open issue of EAS elements concerns spurious *hourglassing instabilities.* This issue was already mentioned in the conclusion of the seminal work on geometrically nonlinear EAS elements by Simo and Armero [131] and has first been thoroughly discussed by Wriggers and Reese [156] for hyperelasticity and by de Souza Neto et al. [37] for elasto-plasticity. One of the most successful attempts to remove the instability has been proposed by Korelc and Wriggers [69] (see also Glaser and Armero [46, 47], who adopted the element for frame invariance). They established that using the *transpose* of the originally employed *Wilson-modes* for the enhanced field removes the instability under compression. Unfortunately though, this attempt could not cure hourglassing under tension. Other methods to suppress the spurious behavior use artificial hourglass stabilization with user-defined parameters (see [6, 46, 47]), hourglass stabilization based on mixed methods (see [123, 124, 147]) or combine the EAS method with other mixed elements (see [7, 8, 31, 52, 63, 64, 102]). However, none of these alternative approaches led to the desired result of a completely stable, locking-free element without artificial stabilization. A final avenue followed to overcome the hourglassing issue is enhancing different kinematic fields than the originally [131] proposed and most frequently employed enhancement of the deformation gradient (F-enhancement). Unfortunately, both enhancement of the Green-Lagrange strain (E-enhancement [17, 20, 41, 46, 66, 93, 147]), which is often employed in shell-structures, and enhancement of the right stretch tensor (U-enhancement [35, 36, 46]) do not lead to entirely stable formulations.
In the present contribution we add another kinematic field to the list of enhanced fields: the spatial displacement gradient (h-enhancement) as suggested by Schmied [125] (PhD-thesis, only available in German). This allows to finally overcome the issue of hourglassing-instabilities of EAS elements both in tension and compression without loosing other favorable properties. On top of that, we establish a hypothesis on why h-enhancement is much more stable. It is closely related to the ad-hoc formulated volume-condition proposed

by Nagtegaal and Fox [97] which is justified in this work. Essentially, we suggest that the instabilities of previously developed EAS elements are linked to the well-known checkerboard modes observed for mixed pressure elements through the eliminated stress field.

In summary, this work introduces a novel PG-EAS element based on h-enhancement which overcomes some of the most serious issues of existing EAS elements. The presentation starts with essential continuum mechanics and the weak form in Section 5.2. The approximations of the various fields with the finite element method are covered in Section 5.3 where focus is put on the Petrov-Galerkin formulation and the required metric shape functions. Analytic solutions of higher-order linear elasticity are given in Appendix 5.A and used for the new volumetric enhanced modes in Section 5.3.2. Section 5.4 covers analytic proofs that the element on the one hand meets basic requirements and on the other hand has interesting features explaining its outstanding performance. The decreased mesh distortion sensitivity and the increased numerical stability are discussed in Sections 5.4.5 and 5.4.9, respectively. Finally, in-depth elastic and elasto-plastic numerical investigations covering a wide range of benchmarks are described in Section 5.5 before conclusions are drawn in Section 5.6.

5.2. Weak form

This section covers fundamental continuum mechanic relations in Section (5.2.1) and introduces the key idea of the novel EAS framework in Section 5.2.2. With these basics at hand we then introduce in Section 5.2.3 the weak form for the new class of EAS elements.

5.2.1. Continuum mechanics

We consider as usual the deformation of a body from its reference configuration $\mathcal{B}_0 \in \mathbb{R}^3$ to a spatial configuration $\mathcal{B} \in \mathbb{R}^3$ (see Figure 5.1). In this process material points $\mathbf{X} \in \mathcal{B}_0$ are mapped to corresponding spatial points $\mathbf{x} = \boldsymbol{\varphi}(\mathbf{X}) \in \mathcal{B}$ by the bijective *deformation map*

$$\boldsymbol{\varphi} \in \mathcal{U} = \left\{\boldsymbol{\varphi} : \mathcal{B}_0 \to \mathbb{R}^3 \,\middle|\, (\boldsymbol{\varphi})_i \in \mathrm{H}_1, \quad \det(\mathrm{D}\boldsymbol{\varphi}) > 0, \quad \boldsymbol{\varphi}(\mathbf{X}) = \bar{\boldsymbol{\varphi}}(\mathbf{X}),\ \mathbf{X} \in \partial_{\varphi}\mathcal{B}_0 \right\}. \quad (5.1)$$

Therein, H_1 denotes a Sobolev space and $\partial_{\varphi}\mathcal{B}_0 \subseteq \partial\mathcal{B}_0$ is a part of the body's boundary $\partial\mathcal{B}_0$ on which the deformation is prescribed by $\bar{\boldsymbol{\varphi}}(\mathbf{X})$. On the remaining part of the boundary $\partial_t\mathcal{B}_0 = \partial\mathcal{B}_0 \setminus \partial_{\varphi}\mathcal{B}_0$ boundary conditions are prescribed in the form of given traction $\mathbf{t}$. Linerization of the deformation map $\boldsymbol{\varphi}$ with respect to the coordinates $\mathbf{X}$ determines the *deformation gradient*[1]

$$\mathbf{F}_{\varphi} = \frac{\partial \boldsymbol{\varphi}}{\partial \mathbf{X}} = \nabla_{\mathbf{X}}\boldsymbol{\varphi}, \quad (5.2)$$

[1] Here and subsequently we frequently omit arguments of functions in favor of a more concise notation.

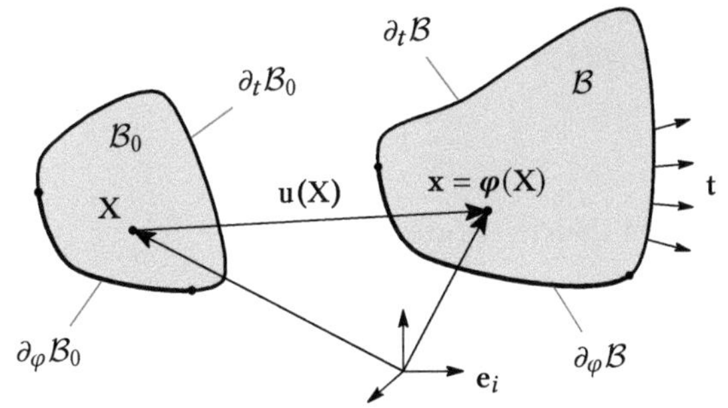

Figure 5.1.: Configurations and kinematics of a deformable body $\mathcal{B}$.

where we use index $(\bullet)_\varphi$ to emphasize that the field $(\bullet)$ depends solely on the deformations (this will not necessarily be the case in subsequent sections). Furthermore,

$$\mathbf{u}(\mathbf{X}) = \boldsymbol{\varphi}(\mathbf{X}) - \mathbf{X} \tag{5.3}$$

is the displacement of a material point $\mathbf{X}$. Taking the derivative of $\mathbf{u}$ with respect to $\mathbf{X}$ and $\mathbf{x}$ yields the material and spatial *displacement gradient* given by

$$\mathbf{H}_\varphi = \frac{\partial \mathbf{u}}{\partial \mathbf{X}} = \nabla_{\mathbf{X}}\mathbf{u} = \mathbf{F}_\varphi - \mathbf{I}, \tag{5.4a}$$

$$\mathbf{h}_\varphi = \frac{\partial \mathbf{u}}{\partial \mathbf{x}} = \nabla_{\mathbf{x}}\mathbf{u} = \mathbf{I} - \mathbf{F}_\varphi^{-1}, \tag{5.4b}$$

respectively. The link of the displacement gradients to the deformation gradient given in (5.4) can easily be established via (5.2) and (5.3). Moreover, in view of (5.4), it is straightforward to establish the relations

$$\mathbf{h} = \mathbf{H}\mathbf{F}^{-1}, \qquad \mathbf{F} = (\mathbf{I} - \mathbf{h})^{-1}, \tag{5.5}$$

frequently used for the present EAS element. We omit index φ in the last equation and subsequently whenever a relation can and will also be used for general and not only purely deformation-based fields.

In addition to the kinematic relations presented above, a constitutive law is required to model different material behavior. A broad class of material models can be written in terms of the *Kirchhoff stress* as

$$\hat{\boldsymbol{\tau}} = \hat{\boldsymbol{\tau}}(\mathbf{F}, \Xi), \tag{5.6}$$

where Ξ denotes the internal variables necessary to describe inelastic behavior. The usual transformation rules

$$\hat{\boldsymbol{\tau}} = \hat{\mathbf{P}}\mathbf{F}^{\mathrm{T}} = \mathbf{F}\hat{\mathbf{S}}\mathbf{F}^{\mathrm{T}} \tag{5.7}$$

relate the constitutive Kirchhoff stress $\hat{\boldsymbol{\tau}}$ to the corresponding *first* and *second Piola-Kirchoff stress* tensors denoted by $\hat{\mathbf{P}} = \hat{\mathbf{P}}(\mathbf{F}, \Xi)$ and $\hat{\mathbf{S}} = \hat{\mathbf{S}}(\mathbf{F}, \Xi)$, respectively.

Remark 5.1. *For a homogeneous hyperelastic material model* (5.6) *is fully defined by a* strain energy function W *and given by*

$$\hat{\tau} = \frac{\partial W}{\partial \mathrm{F}} \mathrm{F}^{\mathrm{T}} = 2\mathrm{F}\frac{\partial W}{\partial \mathrm{C}} \mathrm{F}^{\mathrm{T}}, \tag{5.8}$$

where $\mathrm{C} = \mathrm{F}^{\mathrm{T}}\mathrm{F}$ *is the* right Cauchy-Green *tensor.*

5.2.2. EAS kinematics

The key idea of the EAS method is to *enhance* a kinematic field with an *incompatible part.* This was first proposed for geometrically nonlinear problems by Simo and Armero [131] who recast the total deformation gradient F in the form (F-enhancement)

$$\mathrm{F}(\boldsymbol{\varphi}, \boldsymbol{\alpha}) = \mathrm{F}_{\varphi}(\boldsymbol{\varphi}) + \tilde{\mathrm{F}}(\boldsymbol{\varphi}, \boldsymbol{\alpha}). \tag{5.9}$$

Thus, the compatible (purely deformation dependent) deformation gradient F_{φ} (5.2) is enhanced with the incompatible part $\tilde{\mathrm{F}}$ which contains the additional degrees of freedom $\boldsymbol{\alpha}$. However, as mentioned in the introduction, it is also possible to enhance various other kinematic fields. In the present work we choose to enhance the spatial displacement gradient (5.4b) in the same way as (5.9) such that (h-enhancement)

$$\boxed{\mathrm{h}(\boldsymbol{\varphi}, \boldsymbol{\alpha}) = \mathrm{h}_{\varphi}(\boldsymbol{\varphi}) + \tilde{\mathrm{h}}(\boldsymbol{\varphi}, \boldsymbol{\alpha}),} \tag{5.10}$$

where now instead of $\tilde{\mathrm{F}}$ in (5.9) the *enhanced spatial displacement gradient* $\tilde{\mathrm{h}}$ is the independent field. Note that the total deformation gradient now results from

$$\mathrm{F}(\boldsymbol{\varphi}, \boldsymbol{\alpha}) = (\mathrm{I} - \mathrm{h}(\boldsymbol{\varphi}, \boldsymbol{\alpha}))^{-1} \tag{5.11}$$

which is in analogy to $(5.5)_2$. This idea of enhancing the spatial displacement gradient has to the best knowledge of the authors only been considered in the PhD-Thesis by Schmied [125], which is written in German. In that work it is shown numerically that (5.10) can be used to construct elements without spurious hourglassing in explicit dynamic simulations. In the present contribution we use (5.10) in implicit static problems in order to obtain an hourglassing-free EAS element in both large strain elasticity and elasto-plasticity. Furthermore, we establish a hypothesis why h-enhancement is favorable in terms of stability (see Section 5.4.9).

5.2.3. Weak form

5.2.3.1. Continuous weak form

With the basic continuum mechanic relations at hand we now introduce the weak form of the governing equations for an element based on h-enhancement (5.10). It is given by: Find the deformation $\boldsymbol{\varphi} \in \mathcal{U}$, incompatible displacement gradient $\tilde{\mathbf{h}} \in \mathcal{F} = L_2$ and independent stress tensor $\boldsymbol{\tau} \in \mathcal{T} = L_2$ such that

$$\int_{\mathcal{B}_0} \tilde{\nabla}^s_{\mathbf{x}} \delta\boldsymbol{\varphi} : \hat{\boldsymbol{\tau}} \, \mathrm{d}V + G_{\text{ext}}(\delta\boldsymbol{\varphi}) = 0, \tag{5.12a}$$

$$\int_{\mathcal{B}_0} \delta\tilde{\mathbf{h}} : \left(\hat{\boldsymbol{\tau}} - \boldsymbol{\tau}\right) \mathrm{d}V = 0, \tag{5.12b}$$

$$\int_{\mathcal{B}_0} \delta\boldsymbol{\tau} : \tilde{\mathbf{h}} \, \mathrm{d}V = 0, \tag{5.12c}$$

is satisfied for arbitrary test functions $\delta\boldsymbol{\varphi} \in \mathcal{V} = \{\boldsymbol{\varphi} : \mathcal{B}_0 \to \mathbb{R}^3 | \ (\boldsymbol{\varphi})_i \in \mathrm{H}_1, \ \boldsymbol{\varphi}(\mathbf{X}) = \mathbf{0}, \ \mathbf{X} \in \partial_\varphi \mathcal{B}_0\}$, $\delta\tilde{\mathbf{h}} \in \mathcal{E} = L_2$ and $\delta\boldsymbol{\tau} \in \mathcal{S} = L_2$. Here, $\delta(\bullet)$ is used for test functions and does not denote variations since (5.12) can in general not be derived from a variational functional.[2] Moreover, $\hat{\boldsymbol{\tau}}$ is the constitutive Kirchoff stress (5.6) evaluated with F defined via (5.11) and (5.10). The gradient $\tilde{\nabla}^s_{\mathbf{x}}(\bullet)$ is the symmetric part of the spatial gradient

$$\tilde{\nabla}_{\mathbf{x}}(\bullet) = \nabla_{\mathbf{X}}(\bullet)\mathbf{F}_{\varphi}^{-1}, \tag{5.13}$$

which is closely related to $(5.5)_1$. Finally, the integrals in (5.12) are named internal parts of the weak form and $G_{\text{ext}}(\delta\boldsymbol{\varphi})$ contains the contributions of the prescribed external body forces **b** and surface loads **t**.

Remark 5.2. *It is important to note that, unless the test function $\delta\tilde{\mathbf{h}}$ has the form $\delta\tilde{\mathbf{h}} = \delta\tilde{\mathbf{H}}\mathbf{F}^{-1}$, $\delta\tilde{\mathbf{H}} \in L_2$ in analogy to $(5.5)_1$, the weak form (5.12) leads in general to an unsymmetric stiffness matrix even without the Petrov-Galerkin approach introduced in Section 5.3 (see Schmied [125]). Unfortunately, choosing $\delta\tilde{\mathbf{h}} = \delta\tilde{\mathbf{H}}\mathbf{F}^{-1}$ results in an element which exhibits spurious instabilities. However, the stiffness matrix is already unsymmetric due to the Petrov-Galerkin approach in Section 5.3 such that the unsymmetry imposed by the weak form is not a major issue.*

Remark 5.3. *The proposed weak form (5.12) is by far not the only possibility to develop an EAS framework. E.g., the standard weak form for nonlinear EAS elements with F-enhancement (5.9) (e.g., [47, 104, 131]) is given by*

$$\int_{\mathcal{B}_0} \nabla_{\mathbf{X}} \delta\boldsymbol{\varphi} : \hat{\mathbf{P}} \, \mathrm{d}V + G_{ext}(\delta\boldsymbol{\varphi}) = 0, \tag{5.14a}$$

$$\int_{\mathcal{B}_0} \delta\tilde{\mathbf{F}} : (\hat{\mathbf{P}} - \mathbf{P}) \, \mathrm{d}V = 0, \tag{5.14b}$$

$$\int_{\mathcal{B}_0} \delta\mathbf{P} : \tilde{\mathbf{F}} \, \mathrm{d}V = 0. \tag{5.14c}$$

A similar formulation based on enhancement of the Green-Lagrange strain (E-enhancement) is also possible (e.g., [17, 46, 65, 66, 147]). The crucial advantage of the weak form (5.12) based

[2] The same holds, e.g., also for the EAS element by Glaser and Armero [47] due to the transformation needed for objectivity.

on $\mathbf{h}$ *is that it allows the construction of elements without spurious instabilities. Thus,* (5.12) *is mainly considered in the sequel. Nevertheless, we also cover elements based on* (5.14) *for comparison with the novel displacement gradient enhanced elements.*

5.2.3.2. Discrete weak form

One particularity of EAS elements regarding the weak form is the assumption of L_2-orthogonal discrete[3] test and trial spaces for the independent stress $\boldsymbol{\tau}^{\mathrm{h}}$ and incompatible strain $\tilde{\mathbf{h}}^{\mathrm{h}}$ (see [131, 134] and Section 5.4.7). In that way the independent stress field $\boldsymbol{\tau}^{\mathrm{h}}$ can be completely eliminated from (5.12) and the only remaining unknowns are $\boldsymbol{\varphi} \in \mathcal{U}^{\mathrm{h}}$ and $\tilde{\mathbf{h}}^{\mathrm{h}} \in \mathcal{E}^{\mathrm{h}}$. The discrete version of the weak form (5.12) is then given by

$$
\boxed{
\begin{aligned}
G_{\varphi}^{\mathrm{h}} &= \int_{\mathcal{B}_0^{\mathrm{h}}} \tilde{\nabla}_{\mathrm{x}}^{s} \delta\boldsymbol{\varphi}^{\mathrm{h}} : \hat{\boldsymbol{\tau}}^{\mathrm{h}} \, \mathrm{d}V + G_{\mathrm{ext}}^{\mathrm{h}}(\delta\boldsymbol{\varphi}^{\mathrm{h}}) = 0 \qquad && \forall\, \delta\boldsymbol{\varphi}^{\mathrm{h}} \in \mathcal{V}^{\mathrm{h}}, \\
G_{\alpha}^{\mathrm{h}} &= \int_{\mathcal{B}_0^{\mathrm{h}}} \delta\tilde{\mathbf{h}}^{\mathrm{h}} : \hat{\boldsymbol{\tau}}^{\mathrm{h}} \, \mathrm{d}V = 0 && \forall\, \delta\tilde{\mathbf{h}}^{\mathrm{h}} \in \mathcal{F}^{\mathrm{h}},
\end{aligned}
}
\begin{aligned}
&(5.15\text{a})\\
&(5.15\text{b})
\end{aligned}
$$

where $\mathcal{U}^{\mathrm{h}}$, $\mathcal{V}^{\mathrm{h}}$, $\mathcal{E}^{\mathrm{h}}$ and $\mathcal{F}^{\mathrm{h}}$ are the discrete ansatz spaces for the various fields which are specified in Section 5.3.2.

5.3. Finite element approximation

We use very similar ansatz spaces for the present finite element as proposed by Pfefferkorn and Betsch [106] for a linear *Petrov-Galerkin* EAS (PG-EAS) element. These ansatz spaces are based on three design conditions determined in aforementioned reference and allow the construction of relatively simple, widely applicable, low-order, unsymmetric elements with exceptionally high accuracy in case of distorted meshes. However, the present nonlinear framework requires a few modifications to ensure objectivity and account for the nonlinear kinematics. In this section we focus on these specifics and only briefly summarize the ansatz functions discussed at length by Pfefferkorn and Betsch [106]. Focus is put on three-dimensional (3D) *hexahedral elements* which can easily be reduced to two-dimensional (2D) quadrilaterals.

Moreover, we propose an improved version of the higher order enhanced modes required for 3D problems. A minor modification based on analytic solutions for linear elasticity given in Appendix 5.A circumvents the need for the computationally expensive and intricate orthogonalization procedure originally employed in [106].

[3] Here and subsequently we use superscript $(\bullet)^{\mathrm{h}}$ to denote finite element discreteizations.

5.3.1. Geometry and skew coordinate frame

The reference geometry $\mathbf{X}^{\mathrm{h},e}$ of a hexahedral finite element Ω_0^e with nodes $\mathbf{X}_i^e$, $i = 1, \ldots, 8$ is described as usual with the *isoparametric* map such that the geometry map and the corresponding *Jacobian* are given by

$$\mathbf{X}^{\mathrm{h},e}(\boldsymbol{\xi}) = \sum_{i=1}^{8} N_i(\boldsymbol{\xi})\mathbf{X}_i^e, \qquad \mathbf{J}^{\mathrm{h},e}(\boldsymbol{\xi}) = \frac{\partial \mathbf{X}^{\mathrm{h},e}}{\partial \boldsymbol{\xi}} = \sum_{i=1}^{8} \mathbf{X}_i^e \otimes \nabla_{\boldsymbol{\xi}} N_i, \tag{5.16}$$

respectively. Therein, N_i are the standard *trilinear Lagrangian* shape functions and $\boldsymbol{\xi} \in \hat{\Omega} = [-1, +1]^3$ denotes the *isoparametric coordinates* defined on reference element $\hat{\Omega}$.

Another set of coordinates, which is extensively used for the ansatz spaces described in Section 5.3.2, are the *skew coordinates*. They were first proposed by Yuan et al. [159] and Wisniewski and Turska [150] in the context of assumed stress elements and are given by

$$\bar{\boldsymbol{\xi}} = \begin{bmatrix} \bar{\xi} & \bar{\eta} & \bar{\zeta} \end{bmatrix}^{\mathrm{T}} = \mathbf{J}_0^{-1}\left(\mathbf{X}^{\mathrm{h},e} - \mathbf{X}_0\right), \tag{5.17}$$

where

$$\mathbf{J}_0 = \mathbf{J}^{\mathrm{h},e}(\mathbf{0}), \qquad \mathbf{X}_0 = \mathbf{X}^{\mathrm{h},e}(\mathbf{0}) \tag{5.18}$$

are the Jacobian and position defined in (5.16) evaluated at the element center $\boldsymbol{\xi} = \mathbf{0}$. An alternative form of the skew coordinates which directly links them to the isoparametric coordinates is given by [106]

$$\bar{\boldsymbol{\xi}} = \boldsymbol{\xi} + \sum_{A=1}^{4} \mathbf{c}^A H_A(\boldsymbol{\xi}), \tag{5.19}$$

and is based on an alternative representation of the shape functions (see, e.g., [16, 104, 154]). In the equation above we use

$$\mathbf{c}^A = \mathbf{J}_0^{-1}\frac{1}{8}\sum_{i=1}^{8} \mathbf{X}_i^e h_i^A, \tag{5.20a}$$

$$H_1 = \eta\zeta, \qquad H_2 = \xi\zeta, \qquad H_3 = \xi\eta, \qquad H_4 = \xi\eta\zeta, \tag{5.20b}$$

$$\begin{aligned}
\mathbf{h}^1 &= \begin{bmatrix} +1 & +1 & -1 & -1 & -1 & -1 & +1 & +1 \end{bmatrix}^{\mathrm{T}}, \\
\mathbf{h}^2 &= \begin{bmatrix} +1 & -1 & -1 & +1 & -1 & +1 & +1 & -1 \end{bmatrix}^{\mathrm{T}}, \\
\mathbf{h}^3 &= \begin{bmatrix} +1 & -1 & +1 & -1 & +1 & -1 & +1 & -1 \end{bmatrix}^{\mathrm{T}}, \\
\mathbf{h}^4 &= \begin{bmatrix} -1 & +1 & -1 & +1 & +1 & -1 & +1 & -1 \end{bmatrix}^{\mathrm{T}}.
\end{aligned} \tag{5.20c}$$

The most important properties of the skew coordinate map (5.17) are its frame-indifference (see Section 5.4.2) and the fact that (5.17) is an affine map of the physical coordinates. These features facilitate the construction of proper ansatz spaces. More details on the skew coordinate frame can, e.g., be found in the work of Pfefferkorn and Betsch [106].

5.3.2. Ansatz spaces

5.3.2.1. Test function for the deformation

For the ansatz of the test function of the deformation $\delta\boldsymbol{\varphi}^{\mathrm{h},e} \in \mathcal{V}^{\mathrm{h}}$ we use the standard isoparametric concept such that

$$\delta\boldsymbol{\varphi}^{\mathrm{h},e} = \sum_{i=1}^{8} N_i \delta\boldsymbol{\varphi}_i^e \tag{5.21}$$

is defined in the same way as (5.16) with the same trilinear Lagrangian shape functions N_i and the nodal weights $\delta\boldsymbol{\varphi}_i^e$. For the discrete spatial gradient of $\delta\boldsymbol{\varphi}^{\mathrm{h},e}$ we define

$$\tilde{\nabla}_{\mathbf{x}} \delta\boldsymbol{\varphi}^{\mathrm{h},e} = \nabla_{\mathbf{X}} \delta\boldsymbol{\varphi}^{\mathrm{h},e} \mathbf{F}_0^{-1} \tag{5.22}$$

$$\mathbf{F}_0 = \mathbf{F}_{\varphi}^{\mathrm{h},e}\big|_{\boldsymbol{\xi}=\mathbf{0}}, \tag{5.23}$$

in analogy to (5.13). Therein, $\mathbf{F}_{\varphi}^{\mathrm{h},e}$ denotes the compatible deformation gradient which is subsequently defined in (5.28) and $\mathbf{F}_0$ is its evaluation at the element center $\boldsymbol{\xi} = \mathbf{0}$. Employing $\mathbf{F}_0$ in (5.22) ensures consistency with ordinary continuum mechanics as well as objectivity of the finite element (see Sections 5.4.1 and 5.4.2).

5.3.2.2. Trial function for the deformation

The next field to be discretized is the trial function of the deformation $\boldsymbol{\varphi}^{\mathrm{h},e} \in \mathcal{U}^{\mathrm{h}}$. Similarly to the approximation of the test function (5.21) we use

$$\boldsymbol{\varphi}^{\mathrm{h},e} = \sum_{i=1}^{8} M_i^e \boldsymbol{\varphi}_i^e, \tag{5.24}$$

where $\boldsymbol{\varphi}_i^e$ are the nodal weights. The only difference to the discretization of the test function (5.21) is that (5.24) uses the *metric shape functions* M_i^e instead of the isoparametric N_i. This alternative set of shape functions is constructed elementwise in the physical space[4] using the skew coordinates (5.17). For a hexahedral element Ω_0^e the metric shape functions M_i^e are a linear combination of the eight monomials $1, \bar{\xi}, \bar{\eta}, \bar{\zeta}, \bar{\xi}\bar{\eta}, \bar{\eta}\bar{\zeta}, \bar{\zeta}\bar{\xi}, \bar{\xi}\bar{\eta}\bar{\zeta}$ and given by

$$\begin{bmatrix} M_1^e & \cdots & M_8^e \end{bmatrix} = \begin{bmatrix} 1 & \bar{\xi} & \bar{\eta} & \cdots & \bar{\xi}\bar{\eta}\bar{\zeta} \end{bmatrix} \left(\mathbf{A}_{\bar{\xi}}\right)^{-1}, \quad \mathbf{A}_{\bar{\xi}} = \begin{bmatrix} 1 & \bar{\xi}_1 & \bar{\eta}_1 & \cdots & \bar{\xi}_1\bar{\eta}_1\bar{\zeta}_1 \\ \vdots & & & \ddots & \vdots \\ 1 & \bar{\xi}_8 & \bar{\eta}_8 & \cdots & \bar{\xi}_8\bar{\eta}_8\bar{\zeta}_8 \end{bmatrix}, \tag{5.25}$$

[4] Shape functions constructed in the physical space are in general termed metric shape functions [85].

where $\bar{\xi}_j^e$ are the skew coordinates of the nodes. Relation (5.25) can be established by enforcing the Kronecker-delta property $M_i^e(\bar{\xi}_j^e) = \delta_{ij}$ at all eight nodes [106]. Vice versa, evaluating (5.25) at the nodes yields

$$\begin{bmatrix} M_1^e(\bar{\xi}_1) & \cdots & M_8^e(\bar{\xi}_1) \\ \vdots & \ddots & \vdots \\ M_1^e(\bar{\xi}_8) & \cdots & M_8^e(\bar{\xi}_8) \end{bmatrix} = \mathbf{A}_{\bar{\xi}} \left(\mathbf{A}_{\bar{\xi}}\right)^{-1} = \mathbf{I}_{8\times 8}. \tag{5.26}$$

The derivatives of the metric shape functions (5.25) with respect to the skew coordinates can be computed in a straightforward way from (5.25) and are given by

$$\begin{bmatrix} \nabla_{\bar{\xi}} M_1^e & \cdots & \nabla_{\bar{\xi}} M_8^e \end{bmatrix} = \begin{bmatrix} 0 & 1 & 0 & 0 & \bar{\eta} & 0 & \bar{\zeta} & \bar{\eta}\bar{\zeta} \\ 0 & 0 & 1 & 0 & \bar{\xi} & \bar{\zeta} & 0 & \bar{\xi}\bar{\zeta} \\ 0 & 0 & 0 & 1 & 0 & \bar{\eta} & \bar{\xi} & \bar{\xi}\bar{\eta} \end{bmatrix} \left(\mathbf{A}_{\bar{\xi}}\right)^{-1}. \tag{5.27}$$

The inverse matrix $(\mathbf{A}_{\bar{\xi}})^{-1}$ used in both (5.25) and (5.27) exists as long as there are no coincident nodes and the element's volume does not vanish. Fortunately, the computationally expensive task of evaluating (5.25) and (5.27) for every element has to be conducted only once at problem initialization and the results can be stored for all further Newton iterations and load steps. This is because the skew coordinates (5.17) are defined in terms of the constant reference coordinates rather than the current configuration used, e.g., by Li et al. [76]. Therefore, the increased numerical effort to compute the metric shape functions is hardly relevant in nonlinear simulations.

Further important properties of the metric shape functions are their frame-indifference and isotropy[5] due to the use of the skew coordinates. On top of that M_i^e are a partition of unity $\sum_{i=1}^{8} M_i^e = 1$, have by construction the Kronecker-delta property (5.26) and, in case of a constant Jacobian $\mathbf{J}^{\mathrm{h},e} = \mathbf{J}_0$, the metric shape functions M_i^e coincide with their isoparametric counter parts N_i. Finally and most importantly, their definition in the skew space enables the construction of complete ansatz functions in the physical space (see Section 5.3.2.3).

Returning to (5.24) the metric shape functions (5.25) define the discrete trial function of the deformation and allow to compute corresponding kinematic fields. First, in view of (5.2), the discrete displacement-based deformation gradient can now be written as

$$\mathbf{F}_{\varphi}^{\mathrm{h},e} = \sum_{i=1}^{8} \boldsymbol{\varphi}_i^e \otimes \nabla_{\mathbf{X}} M_i^e, \tag{5.28}$$

where the derivatives of the metric shape functions M_i^e with respect to the reference coordinates $\mathbf{X}$ can easily be determined from (5.27), (5.17) and the chain rule such that

$$\nabla_{\mathbf{X}} M_i^e = \mathbf{J}_0^{-\mathrm{T}} \nabla_{\bar{\xi}} M_i^e. \tag{5.29}$$

[5] Invariance to node numbering, see Section 5.4.2.

Second, with relations (5.28) and (5.5) at hand the discrete compatible part $\mathbf{h}_{\varphi}^{\mathrm{h},e}$ of the displacement gradient (5.10) can be written as

$$\mathbf{h}_{\varphi}^{\mathrm{h},e} = \mathbf{I} - \left(\mathbf{F}_{\varphi}^{\mathrm{h},e}\right)^{-1}. \tag{5.30}$$

5.3.2.3. Trial function for the enhanced strain

The enhanced part $\tilde{\mathbf{h}}^{\mathrm{h},e} \in \mathcal{E}^{\mathrm{h}}$ of the total displacement gradient (5.10) and the corresponding test function (see Section 5.3.2.4) are approximated elementwise which facilitates static condensation of the additional degrees of freedom on element level. To ensure objectivity and to pass the patch test (see Sections 5.4.2 and 5.4.6) we propose the form

$$\tilde{\mathbf{h}}^{\mathrm{h},e} = \bar{\mathbf{h}}^{\mathrm{h},e}(\boldsymbol{\alpha}^e)\,\mathbf{F}_0^{-1}, \tag{5.31}$$

where $\mathbf{F}_0$ denotes the compatible deformation gradient $\mathbf{F}_{\varphi}^{\mathrm{h},e}$ at the element center $\boldsymbol{\xi} = \mathbf{0}$ (see (5.23)) and $\bar{\mathbf{h}}^{\mathrm{h},e} = \bar{\mathbf{h}}_1^{\mathrm{h},e} + \bar{\mathbf{h}}_2^{\mathrm{h},e}$ includes the incompatible degrees of freedom $\boldsymbol{\alpha}^e$ and consists of two parts. The first set of enhanced ansatz functions is motivated by examining (5.25) which reveals that the three monomials $\tilde{m}_1^e = \bar{\xi}^2$, $\tilde{m}_2^e = \bar{\eta}^2$ and $\tilde{m}_3^e = \bar{\zeta}^2$ are missing for a fully quadratic deformation field. However, as shown by Pfefferkorn and Betsch [106], the functions $\tilde{m}_j^e$, $j = 1, 2, 3$ cannot be used directly in the form of incompatible modes. The modification

$$\tilde{M}_j^e = \tilde{m}_j^e(\bar{\xi}) - \sum_{i=1}^{8} M_i^e(\bar{\xi})\tilde{m}_j^e(\bar{\xi}_i), \qquad j = 1, 2, 3 \tag{5.32}$$

is necessary to ensure that the incompatible displacement do not contribute to the nodal deformations, that is, $\tilde{M}_j^e(\bar{\xi}_i) = 0$ holds. Thus, (5.32) ensures that $\boldsymbol{\varphi}_i^e$ introduced in (5.24) are actual nodal deformations. Moreover, the modified functions $\tilde{M}_j^e$ coincide with the bubble functions by Wilson et al. [149] in case of undistorted meshes (see [106]). Ultimately, (5.32) can be used in the sense of incompatible modes for $\bar{\mathbf{h}}^{\mathrm{h},e}$ which yields

$$\bar{\mathbf{h}}_1^{\mathrm{h},e} = \sum_{j=1}^{3} \boldsymbol{\alpha}_j^e \otimes \nabla_{\mathbf{X}} \tilde{M}_j^e \tag{5.33}$$

and has the same structure as the approximation of the deformation gradient in (5.28). The last equation implies that $\bar{\mathbf{h}}_1^{\mathrm{h},e}$ includes three incompatible displacement modes with a total of nine additional degrees of freedom arranged in the vectors $\boldsymbol{\alpha}_j^e$, $j = 1, 2, 3$. The second set of enhanced modes contains three modes and is necessary to avoid a mild form of volumetric locking [132] (see also [106]). For the present finite element we use the modes

$$\bar{\mathbf{h}}_2^{\mathrm{h},e} = \mathbf{J}_0^{-\mathrm{T}} \begin{bmatrix} \alpha_{11}^e \bar{\xi}\bar{\zeta} + \alpha_{12}^e \bar{\xi}\bar{\eta} & 0 & 0 \\ 0 & \alpha_{10}^e \bar{\eta}\bar{\zeta} + \alpha_{12}^e \bar{\xi}\bar{\eta} & 0 \\ 0 & 0 & \alpha_{10}^e \bar{\eta}\bar{\zeta} + \alpha_{11}^e \bar{\xi}\bar{\zeta} \end{bmatrix} \mathbf{J}_0^{-1} \tag{5.34}$$

which are defined in terms of skew coordinates and represent a slight modification of the modes used by Pfefferkorn and Betsch [106]. This modification is inspired by the analytic solutions for linear elastic higher order displacement modes presented in Appendix 5.A and its major advantage is that it allows to circumvent the need for an orthogonalization procedure as is further outlined in Section 5.3.2.4.

5.3.2.4. Test function for the enhanced strain

The last field which has to be discretized is the test function for the enhanced strain $\delta\tilde{\mathbf{h}}^{\mathrm{h},e} \in \mathcal{F}^{\mathrm{h}}$. Here we assume the form

$$\delta\tilde{\mathbf{h}}^{\mathrm{h},e} = \mathbf{F}_0 \delta\bar{\mathbf{h}}^{\mathrm{h},e} \mathbf{F}_0^{\mathrm{T}} \tag{5.35}$$

where again the deformation gradient at the element center (5.23) is employed for objectivity (see Section 5.4.2). Moreover, $\delta\bar{\mathbf{h}}^{\mathrm{h},e}$ is given by

$$\delta\bar{\mathbf{h}}^{\mathrm{h},e} = \frac{1}{j^{\mathrm{h},e}} \mathbf{J}_0^{-\mathrm{T}} \delta\hat{\mathbf{h}}^{\mathrm{h},e} \mathbf{J}_0^{-1}, \tag{5.36}$$

where $j^{\mathrm{h},e} = \det(\mathbf{J}^{\mathrm{h},e})$ and $\mathbf{J}_0$ is defined in (5.18). The last equation includes the transformation from the reference element $\hat{\Omega}$ to the physical configuration Ω_0^e and has the same structure as the test function for the incompatible strain introduced by Pfefferkorn and Betsch [106] (see also Simo and Rifai [134]). In view of the symmetry of $\hat{\boldsymbol{\tau}}^{\mathrm{h},e}$ and the transformations given in (5.35) and (5.36) the skew-symmetric contribution of $\delta\hat{\mathbf{h}}^{\mathrm{h},e}$ is immaterial in (5.15b). For the symmetric part we assume the form

$$\delta\hat{\mathbf{h}}_v^{\mathrm{h},e} = \left[\begin{array}{ccc|cccccc|ccc} \xi & 0 & 0 & 0 & 0 & 0 & 0 & 0 & 0 & 0 & \xi\zeta - \hat{h}_{11}^1 & \xi\eta - \hat{h}_{12}^1 \\ 0 & \eta & 0 & 0 & 0 & 0 & 0 & 0 & 0 & \eta\zeta - \hat{h}_{10}^2 & 0 & \xi\eta - \hat{h}_{12}^2 \\ 0 & 0 & \zeta & 0 & 0 & 0 & 0 & 0 & 0 & \eta\zeta - \hat{h}_{10}^3 & \xi\zeta - \hat{h}_{11}^3 & 0 \\ 0 & 0 & 0 & \xi & \eta & 0 & 0 & 0 & 0 & 0 & 0 & 0 \\ 0 & 0 & 0 & 0 & 0 & \eta & \zeta & 0 & 0 & 0 & 0 & 0 \\ 0 & 0 & 0 & 0 & 0 & 0 & 0 & \xi & \zeta & 0 & 0 & 0 \end{array}\right] \begin{bmatrix} \alpha_1 \\ \alpha_2 \\ \vdots \\ \alpha_{12} \end{bmatrix}, \tag{5.37a}$$

$$\begin{aligned} \hat{h}_{10}^2 &= \tfrac{1}{3}(\xi c_1^1 - \zeta c_3^1), & \hat{h}_{11}^1 &= \tfrac{1}{3}(\eta c_2^2 - \zeta c_3^2), & \hat{h}_{12}^1 &= \tfrac{1}{3}(\zeta c_3^3 - \eta c_2^3), \\ \hat{h}_{10}^3 &= \tfrac{1}{3}(\xi c_1^1 - \eta c_2^1), & \hat{h}_{11}^3 &= \tfrac{1}{3}(\eta c_2^2 - \xi c_1^2), & \hat{h}_{12}^2 &= \tfrac{1}{3}(\zeta c_3^3 - \xi c_1^3), \end{aligned} \tag{5.37b}$$

where $(\bullet)_v$ denotes vector-matrix notation and $\hat{h}_j^i$ are correction terms which ensure orthogonality with the analytic stress modes given subsequently in (5.58). The corrections are computed using components of the vectors $\mathbf{c}^A$ defined in (5.20a) which vanish in case of regular meshes. The modes in the equation above are a modification of the ones proposed by Pfefferkorn and Betsch [106] and motivated by the analytic solutions for linear elasticity presented in Appendix 5.A. Ultimately, the modification allows to skip the time consuming and intricate *orthogonalization procedure* employed in [106] while maintaining the high performance of the element. Thus, using the modified (5.37) is a valuable improvement of the element proposed by Pfefferkorn and Betsch [106].

5.3.2.5. Total displacement and deformation gradient

In view of the approximations (5.30), (5.31) and the kinematic relations (5.10), (5.11) the discrete total deformation and displacement gradient required to compute the constitutive stress can be written as

$$\mathbf{h}^{\mathrm{h},e} = \mathbf{h}_{\varphi}^{\mathrm{h},e} + \tilde{\mathbf{h}}^{\mathrm{h},e}, \tag{5.38}$$

$$\mathbf{F}^{\mathrm{h},e} = \left(\mathbf{I} - \mathbf{h}^{\mathrm{h},e}\right)^{-1}. \tag{5.39}$$

5.3.3. Alternative ansatz spaces

In this section we present a short outline of alternative approaches for the extension of our previous work [106] to the large deformation regime.

5.3.3.1. Enhancement of h

The ansatz spaces presented for the element in Section 5.3.2 are not the only possible choices. For instance, many other transformations to ensure objectivity in (5.22), (5.31) and (5.35) would be possible. Furthermore, there exists a plethora of other ansatz functions for enhanced fields such as, e.g., the transposed Wilson-modes first proposed by Korelc and Wriggers [69].

However, our numerical experiments suggest that the element described in Section 5.3.2 has the most favorable behavior. All other tested elements exhibit spurious instabilities, perform worse with mesh distortion or have other serious drawbacks. In particular, elements which use either $\mathbf{F}^{\mathrm{h},e}$ or $(\mathbf{F}^{\mathrm{h},e})^{-\mathrm{T}}$ instead of $\mathbf{F}_0$ in (5.35) exhibit hourglassing modes similar to the standard EAS element [131]. Replacing $\mathbf{F}_0$ with $(\mathbf{F}_0)^{-\mathrm{T}}$ in (5.35) leads to a form of hourglassing under tension for distorted meshes similar to the one observed in [105] for a special transformation for standard EAS elements. Employing $\mathbf{F}^{\mathrm{h},e}$ instead of $\mathbf{F}_0$ in (5.22) leads to hourglassing under compression for slender elements. Proceeding similarly by replacing $\mathbf{F}_0$ in (5.22) with $\mathbf{F}_{\varphi}^{\mathrm{h},e}$ induces spurious modes under tension in case of instable materials. Finally, using the transposed Wilson-modes leads to worse performance with mesh distortion and also induces hourglassing under tension if there are material instabilities.

5.3.3.2. Enhancement of F

As pointed out in the introduction and Remark 5.3 other fields than $\mathbf{h}$ can be enhanced even though this seems to lead to less well performing elements and in particular hourglassing instabilities. This section briefly summarizes a possible PG-EAS discretization for elements based on enhancement of the deformation gradient $\mathbf{F}$ (5.9) and the corresponding weak form (5.14). To that end we first chose a stress field $\mathbf{P}^{\mathrm{h}}$ which is L_2-orthogonal to the incompatible strain $\tilde{\mathbf{F}}^{\mathrm{h}}$ in order to get a discrete weak form similar to (5.15). Second, the approximations for the test and trial functions of the deformation are chosen exactly the same as presented in Section 5.3.2. Thus, $\delta\boldsymbol{\varphi}^{\mathrm{h},e}$ and $\boldsymbol{\varphi}^{\mathrm{h},e}$ are approximated according to (5.21) and (5.24), respectively. Finally, for the test and trial function of the enhanced deformation gradient suitable choices are

$$\tilde{\mathbf{F}}^{\mathrm{h},e} = \mathbf{F}_{\varphi}^{\mathrm{h},e}\bar{\mathbf{F}}^{\mathrm{h},e}, \tag{5.40a}$$

$$\delta\tilde{\mathbf{F}}^{\mathrm{h},e} = (\mathbf{F}^{\mathrm{h},e})^{-\mathrm{T}}\delta\bar{\mathbf{F}}^{\mathrm{h},e}, \tag{5.40b}$$

where in both equations the first factor ensures objectivity and the second contains the actual ansatz functions for the respective field. In particular, we consider an element that uses the Wilson-modes, that is, (5.33) and (5.36), for $\bar{\mathbf{F}}^{\mathrm{h},e}$ and $\delta\bar{\mathbf{F}}^{\mathrm{h},e}$, respectively. Furthermore, we also test an element that uses the transposed Wilson-modes only for $\bar{\mathbf{F}}^{\mathrm{h},e}$ in the numerical studies in Section 5.5.

5.3.3.3. Enhancement of E

The simplest nonlinear EAS element in terms of finding a suitable discretization is the one based on enhancement of the Green-Lagrange strain **E** and its work-conjugate second Piola-Kirchhoff stress **S**. Since both these fields are defined in the reference configuration no modifications are necessary for objectivity and it is possible to take the ansatz spaces from [106] without adaptions. Unfortunately, these elements suffer from spurious instabilities (see Section 5.5.6).

5.3.4. MIP method for increased robustness

One deficiency of EAS elements is their lack of *robustness* (see Pfefferkorn et al. [109]), by which we mean the high number of load steps and Newton iterations required for convergence in comparison to assumed stress elements. Pfefferkorn et al. [109] extend the *mixed integration point* (MIP) method proposed by Magisano et al. [88] to EAS elements which considerably increases their robustness in many examples. The key idea of that method is to introduce an independent stress tensor $\mathbf{S}_g$ at every Gauss point $g = 1, \ldots, n_{\mathrm{gp}}$. Static condensation of $\mathbf{S}_g$ leads then to a modified element stiffness matrix where the constitutive stress in the geometric parts of the tangent is replaced with the independent stress at the Gauss points.

In the present work we employ the MIP method as presented by Pfefferkorn et al. [109]. We refer to aforementioned reference for details on the implementation (see especially Section 5 in [109]) and our numerical simulation in Section 5.5.5 for the effect of the MIP approach.

5.4. Features and properties of H1U/h12

The element proposed in Sections 5.2 and 5.3 is named **H1U/h12** which stands for a (tri-)linear hexahedral unsymmetric element ("H1U"), with twelve h-enhanced modes ("h12"). In this section we prove that H1U/h12 has many interesting features which lead to the high performance in numerical simulations covered in Section 5.5. The most important features are the improved accuracy in case of distorted meshes discussed in Section 5.4.5 and the increased stability covered in Sections 5.4.8 and 5.4.9. All properties presented in this section are also verified numerically in Section 5.5.

5.4.1. Consistency with ordinary continuum mechanics

In a first step we analyze the mixed weak form (5.12). To ensure that the proposed weak form is a suitable basis for the construction of a finite element we have to show that it is *consistent* with ordinary continuum mechanics. This ensures that a mixed finite element formulation based on (5.12) converges to the analytic results with mesh refinement provided that there are no instabilities, the solution is sufficiently regular and the patch test (see Section 5.4.6) is fulfilled.

To prove consistency of the weak form (5.12) it is sufficient to show that it can be reduced to a purely displacement-based formulation in a continuous (non-discrete) setting. To that end we first obtain from (5.12c) and the standard localization argument $\tilde{\mathbf{h}} = \mathbf{0}$. In view of (5.10), (5.5) and (5.6) we then immediately get $\hat{\boldsymbol{\tau}} = \hat{\boldsymbol{\tau}}_\varphi$ meaning that the constitutive stress does not depend on the additional field $\tilde{\mathbf{h}}$. Inserting this result into (5.12b) reveals $\boldsymbol{\tau} = \hat{\boldsymbol{\tau}}_\varphi$. Finally using $\tilde{\mathbf{h}} = \mathbf{0}$ together with (5.13) and (5.7) in (5.12a) yields

$$\int_{\mathcal{B}_0} \tilde{\nabla}_{\mathbf{X}}^s \delta\boldsymbol{\varphi} : \hat{\boldsymbol{\tau}} \, \mathrm{d}V = \int_{\mathcal{B}_0} \nabla_{\mathbf{X}} \delta\boldsymbol{\varphi} : \hat{\mathbf{P}}_\varphi \, \mathrm{d}V = -G_{\mathrm{ext}}(\delta\boldsymbol{\varphi}), \tag{5.41}$$

which is exactly the same as the standard displacement-based weak form. Thus, the weak form postulated in (5.12) is consistent with ordinary continuum mechanics and a suitable basis for the novel finite element framework.

5.4.2. Frame invariance and objectivity

Next we thoroughly show that the discrete weak form (5.15) and consequently the novel element H1U/h12 is invariant to a global change of reference coordinates $(\bullet)^\flat$ (*frame-invariance*) and a superimposed rigid body motion $(\bullet)^\sharp$ (*objectivity*). The corresponding transformations are

$$\mathbf{X}^\flat = \mathbf{R}\mathbf{X} + \mathbf{c}, \tag{5.42a}$$

$$\boldsymbol{\varphi}^\sharp = \mathbf{Q}\boldsymbol{\varphi} + \mathbf{d}, \tag{5.42b}$$

where $\mathbf{R}, \mathbf{Q} \in SO(3)$ are proper orthogonal tensors and $\mathbf{c}, \mathbf{d} \in \mathbb{R}^3$. First, we consider changes of element geometry described in Section 5.3.1 due to (5.42a) and note that transformation (5.42a) implies nodal coordinate changes according to $\mathbf{X}_i^{e,\flat} = \mathbf{R}\mathbf{X}_i^e + \mathbf{c}$. In view of (5.16) this means $\mathbf{X}^{h,e,\flat} = \mathbf{R}\mathbf{X}^{h,e} + \mathbf{c}$ and $\mathbf{J}^{h,e,\flat} = \mathbf{R}\mathbf{J}^{h,e}$ which yields

$$\bar{\xi}^\flat = (\mathbf{J}_0^\flat)^{-1}\left(\mathbf{X}^{h,e,\flat} - \mathbf{X}_0^\flat\right) = \mathbf{J}_0^{-1}\mathbf{R}^{-1}\mathbf{R}\left(\mathbf{X}^{h,e} - \mathbf{X}_0\right) = \bar{\xi} \tag{5.43}$$

for the skew coordinates (5.17). Thus, the skew coordinates are frame-invariant analogous to the isoparametric frame $\boldsymbol{\xi}^\flat = \boldsymbol{\xi}$ and therefore a suitable basis for the construction of ansatz spaces. Moreover, according to (5.29),

$$(\nabla_{\mathbf{X}}(\bullet))^\flat = \left(\mathbf{J}_0^\flat\right)^{-\mathrm{T}} \nabla_{\bar{\xi}}(\bullet) = \mathbf{R}\nabla_{\mathbf{X}}(\bullet), \qquad (\nabla_{\mathbf{X}}(\bullet))^\flat = \left(\mathbf{J}^\flat\right)^{-\mathrm{T}} \nabla_{\xi}(\bullet) = \mathbf{R}\nabla_{\mathbf{X}}(\bullet), \tag{5.44}$$

holds similarly for fields based on skew and isoparametric coordinates. Second, we consider changes of kinematic fields due to (5.42). By aid of (5.5), (5.42), (5.44) and $\boldsymbol{\varphi}^{h,e,\sharp} = \mathbf{Q}\boldsymbol{\varphi}^{h,e} + \mathbf{d}$ the discrete displacement-based deformation gradient (5.28) and the corresponding displacement gradient (5.30) transform according to

$$\mathbf{F}_\varphi^{h,e,\sharp} = \frac{\partial \boldsymbol{\varphi}^{h,e,\sharp}}{\partial \mathbf{X}^{h,e,\flat}} = \mathbf{Q}\mathbf{F}_\varphi^{h,e}\mathbf{R}^{\mathrm{T}}, \tag{5.45}$$

$$\mathbf{h}_\varphi^{h,e,\sharp} = \mathbf{I} - \left(\mathbf{F}_\varphi^{h,e,\sharp}\right)^{-1} = \mathbf{I} - \mathbf{R}\left(\mathbf{I} - \mathbf{h}_\varphi^{h,e}\right)\mathbf{Q}^{\mathrm{T}}. \tag{5.46}$$

Furthermore, under the assumption that the incompatible degrees of freedom in (5.33) and (5.34) transform via $\boldsymbol{\alpha}_j^{e,\sharp} = \mathbf{R}\boldsymbol{\alpha}_j^e$, $j = 1, 2, 3$ and $\alpha_k^{e,\sharp} = \alpha_k^e$, $k = 10, 11, 12$, respectively, the transformation of the enhanced part of the displacement gradient (5.31) is given by

$$\tilde{\mathbf{h}}^{h,e,\sharp} = \bar{\mathbf{h}}^{h,e,\sharp}\left(\mathbf{F}_0^\sharp\right)^{-1} = \mathbf{R}\bar{\mathbf{h}}^{h,e}\mathbf{R}^{\mathrm{T}}\mathbf{R}\mathbf{F}_0^{-1}\mathbf{Q}^{\mathrm{T}} = \mathbf{R}\tilde{\mathbf{h}}^{h,e}\mathbf{Q}^{\mathrm{T}}. \tag{5.47}$$

Combining transformations (5.46) and (5.47) with (5.38) and (5.39) yields the transformation for the total deformation gradient

$$\mathbf{F}^{h,e,\sharp} = \left(\mathbf{I} - \mathbf{h}_\varphi^{h,e,\sharp} - \tilde{\mathbf{h}}^{h,e,\sharp}\right)^{-1} = \left(\mathbf{R}\left(\mathbf{I} - \mathbf{h}_\varphi^{h,e} - \tilde{\mathbf{h}}^{h,e}\right)\mathbf{Q}^{\mathrm{T}}\right)^{-1} = \mathbf{Q}\mathbf{F}^{h,e}\mathbf{R}^{\mathrm{T}}, \tag{5.48}$$

which is the same as (5.45) and in particular the correct transformation for a deformation gradient (e.g., Ogden [98]). This is only possible because of $\mathbf{F}_0$ included in (5.31) which ensures proper transformation of the enhanced displacement gradient. Third, for a proper isotropic elastic material (5.48) implies that the constitutive Kirchhoff stress tensor transforms according to

$$\hat{\boldsymbol{\tau}}(\mathbf{F}^{h,e,\sharp}) = \hat{\boldsymbol{\tau}}(\mathbf{Q}\mathbf{F}^{h,e}\mathbf{R}^{\mathrm{T}}) = \hat{\boldsymbol{\tau}}(\mathbf{Q}\mathbf{F}^{h,e}) = \mathbf{Q}\hat{\boldsymbol{\tau}}(\mathbf{F}^{h,e})\mathbf{Q}^{\mathrm{T}} \tag{5.49}$$

where the second equality holds because of *isotropy* and the third is due to *material objectivity* (see, e.g., Haupt [50] Ch. 7.2 and 7.3). Now, the only expressions in (5.15) whose transformations have not been established are the test functions. According to (5.21), (5.22), (5.35) and (5.36) we obtain

$$\left(\tilde{\nabla}_{\mathrm{x}}\delta\boldsymbol{\varphi}^{\mathrm{h},e}\right)^{\sharp} = \nabla_{\mathrm{X}}\delta\boldsymbol{\varphi}^{\mathrm{h},e,\sharp}\left(\mathbf{F}_0^{\sharp}\right)^{-1} = \mathbf{Q}\nabla_{\mathrm{X}}\delta\boldsymbol{\varphi}^{\mathrm{h},e}\mathbf{R}^{\mathrm{T}}\mathbf{R}\left(\mathbf{F}_0\right)^{-1}\mathbf{Q}^{\mathrm{T}} = \mathbf{Q}\tilde{\nabla}_{\mathrm{x}}\delta\boldsymbol{\varphi}^{\mathrm{h},e}\mathbf{Q}^{\mathrm{T}}, \tag{5.50}$$

$$\delta\tilde{\mathbf{h}}^{\mathrm{h},e,\sharp} = \mathbf{F}_0^{\sharp}\frac{1}{j^{\mathrm{h},e,\sharp}}\left(\mathbf{J}_0^{\sharp}\right)^{-\mathrm{T}}\delta\hat{\mathbf{h}}^{\mathrm{h},e,\sharp}\left(\mathbf{J}_0^{\sharp}\right)^{-1}\left(\mathbf{F}_0^{\sharp}\right)^{\mathrm{T}} = \mathbf{Q}\delta\tilde{\mathbf{h}}^{\mathrm{h},e}\mathbf{Q}^{\mathrm{T}}, \tag{5.51}$$

by considering the transformations $\delta\boldsymbol{\varphi}^{\mathrm{h},e,\sharp} = \mathbf{Q}\delta\boldsymbol{\varphi}^{\mathrm{h},e} + \mathbf{d}$ and $\delta\bar{\mathbf{h}}^{\mathrm{h},e,\sharp} = \delta\bar{\mathbf{h}}^{\mathrm{h},e}$. In the last two equations the correct transformation is possible due to the proper use of $\mathbf{F}_0$ and $\mathbf{J}_0$ in (5.22), (5.35) and (5.36), respectively.

Finally, substituting (5.49), (5.50) and (5.51) into the internal parts of (5.15) yields

$$G_{\mathrm{int},\varphi}^{\mathrm{h},e,\sharp} = \int_{\Omega_0^{e,\sharp}}\left(\tilde{\nabla}_{\mathrm{x}}^{s}\delta\boldsymbol{\varphi}^{\mathrm{h},e}\right)^{\sharp} : \hat{\boldsymbol{\tau}}^{\mathrm{h},e,\sharp}\,\mathrm{d}V^{\sharp} = \int_{\Omega_0^{e}}\left(\tilde{\nabla}_{\mathrm{x}}^{s}\delta\boldsymbol{\varphi}^{\mathrm{h},e}\right) : \hat{\boldsymbol{\tau}}^{\mathrm{h},e}\,\mathrm{d}V = G_{\mathrm{int},\varphi}^{\mathrm{h},e}, \tag{5.52a}$$

$$G_{\alpha}^{\mathrm{h},e,\sharp} = \int_{\Omega_0^{e,\sharp}}\delta\tilde{\mathbf{h}}^{\mathrm{h},e,\sharp} : \hat{\boldsymbol{\tau}}^{\mathrm{h},e,\sharp}\,\mathrm{d}V^{\sharp} = \int_{\Omega_0^{e}}\delta\tilde{\mathbf{h}}^{\mathrm{h},e} : \hat{\boldsymbol{\tau}}^{\mathrm{h},e}\,\mathrm{d}V = G_{\alpha}^{\mathrm{h},e}, \tag{5.52b}$$

which establishes the frame-invariance and objectivity of the novel element H1U/h12 provided that the external loads transform appropriately. Numerical verifications of this proof can be found in Sections 5.5.1.2 and 5.5.1.3.

5.4.3. Isotropy and path independence

Two further crucial properties for any finite element are *isotropy*, implying invariance to node numbering, and *path-independence* in the case of elastic materials.

Isotropy of H1U/h12 can easily be verified by observing that there are no preferred directions in the ansatz spaces chosen in Section 5.3.2. Thus, H1U/h12 is invariant to node numbering which is shown numerically in Section 5.5.1.2. Examples for anisotropic finite elements are the unsymmetric displacement-based element by Ooi et al. [100] (see also Xie et al. [158]) and the EAS element proposed by Korelc et al. [67]. The former is anisotropic because the rotation of the local element coordinates relies only on a subset of nodes and the second is anisotropic because the volumetric enhanced modes have preferred directions.

Another crucial property is *path independence*, i.e., that the element does not depend on the deformation history in case of a path-independent material model. As most elements, the current H1U/h12 is path-independent since all ansatz functions are defined with respect to the reference configuration. If they depend on the current configuration the element is likely to be path-dependent (e.g., the elements by Li et al. [76, 77]).

5.4.4. Linearized element

In this section the novel element H1U/h12 is *linearized* and it is shown that its linearization coincides with the highly accurate element proposed by Pfefferkorn and Betsch [106]. The only difference concerns the improved volumetric enhanced modes (5.34) and (5.37). Thus, all features of that element such as the much increased coarse mesh accuracy, equivalence to standard EAS elements in case of regular meshes, exact numerical integration of the stiffness matrix by standard Gauss quadrature and, most importantly, the exact solution of many bending problems (see also Section 5.4.5) carry over to H1U/h12 in case of small deformations.

The weak form (5.15) is a functional $f : (\boldsymbol{\varphi}, \boldsymbol{\alpha}) \mapsto f(\boldsymbol{\varphi}, \boldsymbol{\alpha})$ of the deformations $\boldsymbol{\varphi}$ and enhanced degrees of freedom $\boldsymbol{\alpha}$. Its linearization in the reference configuration $(\boldsymbol{\varphi}, \boldsymbol{\alpha}) = (\mathbf{X}, \mathbf{0})$ is given by

$$\mathrm{Lin}_0 f = f(\mathbf{X}, \mathbf{0}) + \Delta_0 f, \qquad \Delta_0 f = \frac{\mathrm{d}}{\mathrm{d}\varepsilon} f(\mathbf{X} + \varepsilon\Delta\boldsymbol{\varphi}, \varepsilon\Delta\boldsymbol{\alpha})\bigg|_{\varepsilon=0}, \tag{5.53}$$

where $\Delta\boldsymbol{\varphi}$, $\Delta\boldsymbol{\alpha}$ are increments of the deformation $\boldsymbol{\varphi}$ and enhanced degrees of freedom $\boldsymbol{\alpha}$, respectively. These increments are discretized in the same way as $\boldsymbol{\varphi}$ and $\boldsymbol{\alpha}$ in (5.21), (5.33) and (5.34). Moreover, operator $\Delta_0(\bullet)$ in (5.53) denotes the Gateaux-derivative of $(\bullet)$ with respect to the increments $\Delta\boldsymbol{\varphi}$, $\Delta\boldsymbol{\alpha}$ evaluated in the reference configuration.

With (5.30), (5.28), (5.31), (5.33) and relation $\Delta\mathbf{A}^{-1} = -\mathbf{A}^{-1}\Delta\mathbf{A}\mathbf{A}^{-1}$, which holds for the linearization of a arbitrary tensors $\mathbf{A}$ (see, e.g., Holzapfel [53]), it is then straightforward to obtain

$$\Delta_0\mathbf{h}_{\varphi}^{\mathrm{h},e} = \mathbf{I}^{-1}\Delta_0\mathbf{F}_{\varphi}^{\mathrm{h},e}\mathbf{I}^{-1} \qquad = \nabla_{\mathbf{X}}\Delta\boldsymbol{\varphi}^{\mathrm{h},e}, \tag{5.54a}$$

$$\Delta_0\tilde{\mathbf{h}}^{\mathrm{h},e} = \Delta_0\bar{\mathbf{h}}^{\mathrm{h},e}\mathbf{I} + 0\Delta_0\mathbf{F}_0^{-1} = \bar{\mathbf{h}}^{\mathrm{h},e}(\Delta\boldsymbol{\alpha}^{\mathrm{h},e}) = \bar{\mathbf{h}}_{\Delta\boldsymbol{\alpha}}^{\mathrm{h},e}, \tag{5.54b}$$

for the compatible and incompatible part of the discrete displacement gradient $\mathbf{h}^{\mathrm{h},e}$, respectively. With this information at hand, linearization of the total deformation gradient (5.39) yields

$$\Delta_0\mathbf{F}^{\mathrm{h},e} = \nabla_{\mathbf{X}}\Delta\boldsymbol{\varphi}^{\mathrm{h},e} + \bar{\mathbf{h}}_{\Delta\boldsymbol{\alpha}}^{\mathrm{h},e}. \tag{5.55}$$

Furthermore, linearizing the constitutive stress (5.6) under the assumption of an elastic material and by aid of (5.55) determines

$$\Delta_0\hat{\boldsymbol{\tau}}^{\mathrm{h},e} = \hat{\mathbb{c}}^{\mathrm{h},e}(\mathbf{I}) : \Delta_0\left(\mathbf{F}^{\mathrm{h},e}\left(\mathbf{F}^{\mathrm{h},e}\right)^{\mathrm{T}}\right) = \hat{\mathbb{C}}_{\mathrm{lin}} : \left(\nabla_{\mathbf{X}}^{s}\Delta\boldsymbol{\varphi}^{\mathrm{h},e} + \mathrm{sym}(\bar{\mathbf{h}}_{\Delta\boldsymbol{\alpha}}^{\mathrm{h},e})\right), \tag{5.56}$$

where $\mathrm{sym}(\bar{\mathbf{h}}_{\Delta\boldsymbol{\alpha}}^{\mathrm{h},e})$ denotes the symmetric part of $\bar{\mathbf{h}}_{\Delta\boldsymbol{\alpha}}^{\mathrm{h},e}$ and $\hat{\mathbb{c}}^{\mathrm{h},e}(\mathbf{F}^{\mathrm{h},e})$ is the constitutive spatial elasticity tensor with the corresponding linear elasticity tensor $\hat{\mathbb{C}}_{\mathrm{lin}} = \hat{\mathbb{c}}^{\mathrm{h},e}(\mathbf{I})$.

Finally, with the auxiliary results above, the linearization (5.53) of the weak form (5.15) is given by

$$\operatorname{Lin}_0 G_{\varphi}^{\mathrm{h}} = \int_{\mathcal{B}_0^h} \nabla_{\mathrm{X}}^{s} \delta\boldsymbol{\varphi}^{\mathrm{h}} : \hat{\mathbb{C}}_{\mathrm{lin}} : \left(\nabla_{\mathrm{X}}^{s} \Delta\boldsymbol{\varphi}^{\mathrm{h}} + \operatorname{sym}(\bar{\mathbf{h}}_{\Delta\boldsymbol{\alpha}}^{\mathrm{h},e})\right) \mathrm{d}V - G_{\mathrm{ext}}^{\mathrm{h}}(\delta\boldsymbol{\varphi}^{\mathrm{h}}) = 0, \tag{5.57a}$$

$$\operatorname{Lin}_0 G_{\alpha}^{\mathrm{h}} = \int_{\mathcal{B}_0^h} \delta\bar{\mathbf{h}}^{\mathrm{h}} : \hat{\mathbb{C}}_{\mathrm{lin}} : \left(\nabla_{\mathrm{X}}^{s} \Delta\boldsymbol{\varphi}^{\mathrm{h}} + \operatorname{sym}(\bar{\mathbf{h}}_{\Delta\boldsymbol{\alpha}}^{\mathrm{h},e})\right) \mathrm{d}V = 0, \tag{5.57b}$$

which is the same as the weak form of the standard linear elastic EAS element. Furthermore, the approximation of the fields in Section 5.3.2 is, apart from the two minor modifications (5.34) and (5.37), taken directly from Pfefferkorn and Betsch [106]. Thus, the only differences of the linearized H1U/h12 in comparison to the linear elastic element in aforementioned reference are (5.34) and (5.37). The linearization of the corresponding 2D element is even exactly the same as the element proposed in the work [106].

5.4.5. Insensitivity to mesh distortion

This section covers the *design conditions* established by Pfefferkorn and Betsch [106] which allow to construct elements with optimal or close to optimal performance *regardless of mesh distortion*. To that end we first briefly repeat the linear elastic case from Reference [106] and then generalize the concept to nonlinear problems.

5.4.5.1. Linear elasticity

For linear elasticity the investigations by MacNeal [84] (see also [106]) show that an element with N degrees of freedom can at best be exact for N displacement modes provided that its stiffness matrix is unsymmetric. In case of a PG-EAS framework Pfefferkorn and Betsch [106] examined the weak form (5.57) and obtained *three design conditions* required to get an exact finite element solution for a specific displacement mode regardless of mesh distortion. The design conditions are:

C1 The test functions for the displacement $\delta\mathbf{u}^{\mathrm{h}}$ have to fulfill the inter-element continuity. This ensures that nodal equilibrium is fulfilled and that the correct solutions for a single element can be generalized to larger patches of elements.

C2 The ansatz spaces for the displacement $\mathbf{u}^{\mathrm{h},e}$ and incompatible strain $\tilde{\boldsymbol{\varepsilon}}^{\mathrm{h},e}$ must be chosen such that resulting $\hat{\boldsymbol{\sigma}}^{\mathrm{h},e}$ includes the analytic stress $\boldsymbol{\sigma}^*$ under the premise of nodally exact displacement.

C3 The test function of the incompatible strain $\delta\tilde{\boldsymbol{\varepsilon}}^{\mathrm{h},e}$ must be L_2-orthogonal to the analytic stress $\boldsymbol{\sigma}^*$. This condition is an extension of the patch test condition for EAS elements originally proposed by Simo and Rifai [134].

These conditions can be used to construct an optimal or close to optimal finite element by choosing a proper set of N analytic modes for which the element is then designed to be exact [106].

Similar to Pfefferkorn and Betsch [106] we use the stress modes usually employed for assumed stress elements (see, e.g., [110, 140, 150]) as analytic modes. They are defined in terms of skew coordinates (5.17) and given by

$$\boldsymbol{\sigma}^* = \mathbf{J}_0 \bar{\boldsymbol{\sigma}}^* \mathbf{J}_0^{\mathrm{T}}, \tag{5.58a}$$

$$\bar{\boldsymbol{\sigma}}_v^*(\bar{\boldsymbol{\xi}}) := \left[\begin{array}{cccccc|cccccc|ccc} 1 & 0 & 0 & 0 & 0 & 0 & \bar{\eta} & \bar{\zeta} & 0 & 0 & 0 & 0 & 0 & 0 & 0 \\ 0 & 1 & 0 & 0 & 0 & 0 & 0 & 0 & \bar{\zeta} & \bar{\xi} & 0 & 0 & 0 & 0 & 0 \\ 0 & 0 & 1 & 0 & 0 & 0 & 0 & 0 & 0 & 0 & \bar{\xi} & \bar{\eta} & 0 & 0 & 0 \\ 0 & 0 & 0 & 1 & 0 & 0 & 0 & 0 & 0 & 0 & 0 & 0 & \bar{\zeta} & 0 & 0 \\ 0 & 0 & 0 & 0 & 1 & 0 & 0 & 0 & 0 & 0 & 0 & 0 & 0 & \bar{\xi} & 0 \\ 0 & 0 & 0 & 0 & 0 & 1 & 0 & 0 & 0 & 0 & 0 & 0 & 0 & 0 & \bar{\eta} \end{array}\right] \begin{bmatrix} \beta_1 \\ \beta_2 \\ \vdots \\ \beta_{15} \end{bmatrix}, \tag{5.58b}$$

where $(\bullet)_v$ denotes a vector notation of the corresponding tensor quantity. Compared to the modes given in [106], (5.58b) does not include the bilinear stress modes which are of subordinate importance and incompatible as pointed out in aforementioned reference. The major advantage of skipping the higher order stress modes is that it allows to straightforwardly construct the L_2-orthogonal enhanced strain field given in (5.37). Thus, Condition C3 can be fulfilled without tedious orthogonalization procedure. Furthermore, (5.58) still includes the patch-test (β_1 to β_6), bending (β_7 to β_{12}) and torsion modes (β_{13} to β_{15}) crucial for many engineering problems (see also Jabareen and Rubin [62]).

Fortunately, element H1U/h12 fulfills all three design Conditions C1 to C3 for the 15 modes (5.58) in linear elasticity without further modification. This can be concluded from the fact that the linearization of H1U/h12 coincides with the element proposed by Pfefferkorn and Betsch [106] (see Section 5.4.4) and the fact that the latter element is specifically designed to fulfill the design conditions for (5.58). Thus, H1U/h12 is exact for (5.58) in linear elasticity.

5.4.5.2. Nonlinear problems

Unfortunately, it is hardly possible to find similar design conditions and analytic modes in the nonlinear case due to several reasons. First, analytic solutions for nonlinear higher order displacement modes scarcely exist. In special cases where analytic solutions can be found as, e.g., for a bending problem (see Ogden [98] Ch. 5.2.4) the solutions are intricate and include non-polynomial functions. However, using such non-polynomial functions for ansatz spaces is no good idea since this prohibits construction of complete spaces. Moreover, due to the Gauss quadrature, there is, e.g., no difference between a suitable cosine function and the corresponding quadratic polynomial ansatz. This means that using such a function has no effect unless undesirable higher order numerical integration is employed. The second difficulty in finding analytic modes for nonlinear problems concerns the plethora of different material models. Even in case of hyperelasticity many models exist (see, e.g., Hartmann and Neff [49]) which makes it almost impossible to fulfill the design conditions for arbitrary material models.

Therefore, we consider a different approach in the present work:

- The element is designed such that it is exact for the rigid body and patch test modes (i.e., β_1 to β_6 in (5.58)) *regardless* of magnitude of the deformation. This is a standard requirement[6] for finite elements and corresponding proofs for H1U/h12 are given in Sections 5.4.2 and 5.4.6, respectively.
- For the bending and torsion modes (i.e., β_7 to β_{15} in (5.58)) exact solutions for the full nonlinear problem are in general not possible due to the reasons listed above. Instead we chose to fulfill the design Conditions C1 to C3 for these modes only for small (linearized) deformations. Fortunately, in view of the linerization in Section 5.4.5.1 H1U/h12 automatically meets this relaxed form of the design condition for nonlinear bending and torsion modes. Therefore, even though we cannot expect exact solutions in large deformation problems, results will still be improved compared to the standard Bubnov-Galerkin approach, at least for small deformations. In fact, the numerical results in Sections 5.5.2 to 5.5.4 show that the accuracy is greatly increased even for highly nonlinear bending.

5.4.6. Patch test

In this Section we show that H1U/h12 passes the patch test, that is, it is capable of correctly representing states of constant stress. This is an important requirement for any finite element and ensures, among other things, convergence with mesh refinement [162]. We consider a homogeneous deformation $\boldsymbol{\varphi}_{\text{ho}} = \mathbf{F}_{\text{ho}}\mathbf{X}+\mathbf{c}_{\text{ho}}$ with constant deformation gradient $\nabla_{\mathbf{X}}\boldsymbol{\varphi} = \mathbf{F}_{\text{ho}}$ and constant vector $\mathbf{c}_{\text{ho}}$. To prove satisfaction of the patch test we now show that $\boldsymbol{\varphi}^{\text{h},e} = \boldsymbol{\varphi}_{\text{ho}}$ is a solution[7] of the numeric problem (5.15).

First, we obtain from the homogeneous deformation $\boldsymbol{\varphi}^{\text{h},e} = \boldsymbol{\varphi}_{\text{ho}}$, (5.28) and (5.23) that $\mathbf{F}_{\varphi}^{\text{h},e} = \mathbf{F}_0 = \mathbf{F}_{\text{ho}}$ since the ansatz for the trial function of the deformations (5.24) includes complete linear polynomials in the physical space. This follows from the affine map (5.17) and (5.25). Second, the non-constant[8] fields (5.33) and (5.34) imply $\tilde{\mathbf{h}}^{\text{h},e} = \mathbf{0}$. Consequently, (5.39) and (5.6) determine

$$\mathbf{F}^{\text{h},e} = \mathbf{F}_{\text{ho}} \qquad \Rightarrow \qquad \hat{\boldsymbol{\tau}}(\mathbf{F}^{\text{h},e}) = \hat{\boldsymbol{\tau}}_{\text{ho}} \tag{5.59}$$

[6] For various nonlinear mixed elements which meet this requirement see, e.g., [5, 47, 131, 132, 135, 145, 152].
[7] It is not necessarily the only solution due to possible instabilities but it suffices to show that $\boldsymbol{\varphi}^{\text{h},e} = \boldsymbol{\varphi}_{\text{ho}}$ is *a* solution to prove that the patch test is fulfilled.
[8] See also Simo and Rifai's [134] design imperative for EAS elements on non-overlapping ansatz spaces of the actual displacement and the enhanced field.

where $\hat{\boldsymbol{\tau}}_{\text{ho}}$ is the constant Kirchhoff stress corresponding to $\mathbf{F}_{\text{ho}}$. These results in combination with the approximations (5.35) and (5.36) allow to recast the second equation of the weak form (5.15b) in the form

$$\int_{\Omega_0^e} \frac{1}{j^{\text{h},e}} \mathbf{F}_0 \mathbf{J}_0^{-\text{T}} \delta\hat{\mathbf{h}}^{\text{h},e} \mathbf{J}_0^{-1} \mathbf{F}_0 j^{\text{h},e} \, \mathrm{d}\hat{\Omega} = \mathbf{0} \quad \Leftrightarrow \quad \int_{\hat{\Omega}} \delta\hat{\mathbf{h}}^{\text{h},e} \, \mathrm{d}\hat{\Omega} = \mathbf{0} \tag{5.60}$$

which is satisfied exactly by (5.37). Moreover, with the first Piola Kirchhoff stress tensor $\hat{\mathbf{P}}_{\text{ho}} = \hat{\boldsymbol{\tau}}_{\text{ho}} \mathbf{F}_{\text{ho}}^{-\text{T}}$ the first equation of the weak form (5.15a) can be written as

$$\int_{\mathcal{B}_0^{\text{h}}} \tilde{\nabla}_{\mathbf{x}}^s \delta\boldsymbol{\varphi}^{\text{h}} : \hat{\boldsymbol{\tau}}_{\text{ho}} \, \mathrm{d}V + G_{\text{ext}}^{\text{h}}(\delta\boldsymbol{\varphi}^{\text{h}}) = 0 \quad \Leftrightarrow \quad \int_{\mathcal{B}_0^{\text{h}}} \nabla_{\mathbf{X}} \delta\boldsymbol{\varphi}^{\text{h}} : \mathbf{P}_{\text{ho}} \, \mathrm{d}V + G_{\text{ext}}^{\text{h}}(\delta\boldsymbol{\varphi}^{\text{h}}) = 0 \tag{5.61}$$

which is the same as for an isoparametric displacement-based finite element well-known to pass the patch test (see, e.g., Zienkiewicz et al. [162]). Thus, H1U/h12 fulfills the patch test since both equations of (5.15) are exactly satisfied for $\boldsymbol{\varphi}^{\text{h},e} = \boldsymbol{\varphi}_{\text{ho}}$. This result is confirmed numerically in Section 5.5.1.1.

5.4.7. L_2-orthogonal discrete stress

As usual for EAS elements, the discrete independent stress fields $\delta\boldsymbol{\tau}^{\text{h},e} \in \mathcal{S}^{\text{h}}$ and $\boldsymbol{\tau}^{\text{h},e} \in \mathcal{T}^{\text{h}}$ have been eliminated from the weak form in Section 5.2.3.2 via *assumed* L_2-orthogonality between the stress and enhanced strain field. In this section we present suitable ansatz spaces for the stress which actually fulfill the assumed orthogonality. These functions enable variationally consistent stress recovery during post processing and play a crucial role in Section 5.4.8 and 5.4.9. Suitable stress approximations are given by

$$\delta\boldsymbol{\tau}^{\text{h},e} = \frac{1}{j^{\text{h},e}} \mathbf{F}_0 \mathbf{J}_0 \delta\bar{\boldsymbol{\tau}}^{\text{h},e} \mathbf{J}_0^{\text{T}} \mathbf{F}_0^{\text{T}} \qquad \in \mathcal{S}^{\text{h}}, \tag{5.62a}$$

$$\boldsymbol{\tau}^{\text{h},e} = \mathbf{F}_0^{-\text{T}} \mathbf{J}_0 \bar{\boldsymbol{\tau}}^{\text{h},e} \mathbf{J}_0^{\text{T}} \mathbf{F}_0^{-1} \qquad \in \mathcal{T}^{\text{h}}, \tag{5.62b}$$

where $j^{\text{h},e} = \det(\mathbf{J}^{\text{h},e})$ and $\mathbf{F}_0, \mathbf{J}_0$ are given in (5.23) and (5.18), respectively. By choosing $\bar{\boldsymbol{\tau}}^{\text{h},e} = \bar{\boldsymbol{\sigma}}^*$ given in (5.58) the independent stress $\boldsymbol{\tau}^{\text{h},e}$ can immediately be eliminated from (5.12b). This can easily be verified with (5.62b), (5.35), (5.36) and the fact that

$$\int_{\Omega_0^e} \delta\tilde{\mathbf{h}}^{\text{h},e} : \boldsymbol{\tau}^{\text{h},e} \, \mathrm{d}V = \int_{\hat{\Omega}} \delta\hat{\mathbf{h}}^{\text{h},e} : \bar{\boldsymbol{\sigma}}^* \, \mathrm{d}\hat{\Omega} = 0 \tag{5.63}$$

holds regardless of element geometry as consequence of the particular choice (5.36) (see also the design Condition C3 in Section 5.4.5.1). A similar procedure can also be applied for the test function of the independent stress in (5.12c). Here we choose $\delta\bar{\boldsymbol{\tau}}^{\text{h},e} = \delta\bar{\boldsymbol{\sigma}}^*$ where $\delta\bar{\boldsymbol{\sigma}}^*$ has the same form as $\bar{\boldsymbol{\sigma}}^*$ given in (5.58) but uses the isoparametric coordinates

ξ instead of the skew $\bar{\xi}$. Unfortunately, the resulting $\delta\boldsymbol{\tau}^{\mathrm{h},e}$ is only automatically L_2-orthogonal to $\tilde{\mathbf{h}}^{\mathrm{h},e}$ given in (5.31) in case of regular meshes.[9] However, for distorted geometries a Gram-Schmidt orthogonalization procedure similar to the one proposed by Pfefferkorn and Betsch [106] allows to straightforwardly construct an orthogonal $\delta\boldsymbol{\tau}^{\mathrm{h},e}$ field. Thus, the test function $\delta\boldsymbol{\tau}^{\mathrm{h},e}$ can always be eliminated from (5.12c) which concludes the proof that suitable L_2-orthogonal stress approximations exist.

5.4.8. L_2-orthogonality to constant pressure

Nagtegaal and Fox [97] proposed the *ad hoc* condition that small changes of the enhanced field may not contribute volume changes. They suggest that this condition should be added to the standard requirements on enhanced fields since they suppose it improves stability of EAS elements. Alternatively, the volume condition can also be obtained by requiring that the test function for the enhanced field is L_2-orthogonal to a *piecewise constant pressure field* (see [97, 125]). We substantiate the claim of increased stability and discuss its connection to the L_2-orthogonality to a piecewise constant pressure field in Section 5.4.9.

Here, we present, motivated by the discussion in Section 5.4.9, a slightly modified version of the condition proposed by Nagtegaal and Fox [97]. Instead of requiring orthogonality to a piecewise constant *constitutive* pressure field we consider L_2-orthogonality between the enhanced fields $\delta\tilde{\mathbf{h}}^{\mathrm{h},e}$, $\tilde{\mathbf{h}}^{\mathrm{h},e}$ and *independent* Kirchhoff stress (see also Schmied [125]) of the form $\delta\boldsymbol{\tau}_p^{\mathrm{h},e} = (j^{\mathrm{h},e})^{-1}\delta p_0^{\mathrm{h},e}\mathbf{I}$ and $\boldsymbol{\tau}_p^{\mathrm{h},e} = p_0^{\mathrm{h},e}\mathbf{I}$ where $\delta p_0^{\mathrm{h},e}$, $p_0^{\mathrm{h},e}$ are piecewise (elementwise) constant.[10] The L_2-orthogonality conditions can now be written as

$$\int_{\Omega_0^e} \delta\tilde{\mathbf{h}}^{\mathrm{h},e} : p_0^{\mathrm{h},e}\mathbf{I}\,\mathrm{d}V = 0, \tag{5.64a}$$

$$\int_{\Omega_0^e} \frac{1}{j^{\mathrm{h},e}}\delta p_0^{\mathrm{h},e}\mathbf{I} : \tilde{\mathbf{h}}^{\mathrm{h},e}\,\mathrm{d}V = 0. \tag{5.64b}$$

Since the stress $\boldsymbol{\tau}_p^{\mathrm{h},e} \in \mathcal{T}^{\mathrm{h}}$ and $\mathcal{T}^{\mathrm{h}}$ includes constant modes (see Section 5.4.7), condition (5.64a) is automatically fulfilled due to the L_2-orthogonality described in Section 5.4.7. Unfortunately, pressure fields $\delta\boldsymbol{\tau}_p^{\mathrm{h},e}$ are in general not part of $\mathcal{S}^{\mathrm{h}}$ due to the orthogonalization procedure required in case of distorted meshes. However, (5.64b) holds automatically in case of regular meshes and seems to be important at least in this relaxed sense.

9 For regular meshes with constant Jacobian $\mathbf{J}^{\mathrm{h},e} = \mathbf{J}_0$ the fields are automatically orthogonal. Otherwise, the correction for no nodal contribution of the enhanced modes in (5.33) and the bilinear modes in (5.34) induce a non-orthogonality.

10 The volumetric-deviatoric split of a Cauchy stress tensor $\boldsymbol{\sigma} = \operatorname{dev}(\boldsymbol{\sigma}) - p^{\mathrm{h},e}\mathbf{I}$ and the transformation $\boldsymbol{\tau} = \det(\mathbf{F})\boldsymbol{\sigma}$ imply that $p_0^{\mathrm{h},e}$ is directly related to piecewise constant "physical" pressure $p^{\mathrm{h},e}$. This holds because $\det(\mathbf{F})$ is also piecewise constant due to the fact that the pressure is usually a function of $\det(\mathbf{F})$ alone.

Remark 5.4. *For comparison, considering an F-enhanced EAS element,* (5.14) *and the transformation* $\mathbf{P}_p^{h,e} = \boldsymbol{\tau}_p^{h,e}(\mathbf{F}^{h,e})^{-\mathrm{T}}$ *yields*

$$\int_{\Omega_0^e} \delta\tilde{\mathbf{F}}^{h,e} : \mathbf{P}_p^{h,e} \, \mathrm{d}V = 0 \qquad \Leftrightarrow \qquad \int_{\Omega_0^e} \mathrm{tr}\left(\delta\tilde{\mathbf{F}}^{h,e}(\mathbf{F}^{h,e})^{-1}\right) \mathrm{d}V = 0 \tag{5.65a}$$

instead of (5.64a). *This is a nonlinear condition which cannot easily be fulfilled for arbitrary deformation gradients* $\mathbf{F}^{h,e}$ *(see also Nagtegaal and Fox [97]). A possibility is choosing* $\delta\tilde{\mathbf{F}}^{h,e} = \delta\tilde{\mathbf{h}}^{h,e}\mathbf{F}^{h,e}$ *for the test function of the enhanced deformation gradient. Unfortunately, this is not enough to get a stable finite element as our numerical experiments confirmed. Apparently, it would be necessary to simultaneously fulfill*

$$\int_{\Omega_0^e} \delta\boldsymbol{\tau}_p^{h,e}(\mathbf{F}^{h,e})^{-\mathrm{T}} : \tilde{\mathbf{F}}^{h,e} \, \mathrm{d}V = 0 \qquad \Leftrightarrow \qquad \int_{\Omega_0^e} \mathrm{tr}\left(\tilde{\mathbf{F}}^{h,e}(\mathbf{F}^{h,e})^{-1}\right) \mathrm{d}V = 0 \tag{5.65b}$$

at least for regular meshes. The last equation corresponds to (5.64b) *and in contrast to* (5.65a) *it seems hardly possible to find suitable ansatz functions for* $\tilde{\mathbf{F}}^{h,e}$ *such that* (5.65b) *is satisfied for all* $\mathbf{F}^{h,e}$ *since both* $\tilde{\mathbf{F}}^{h,e}$ *and* $\mathbf{F}^{h,e}$ *depend on* $\tilde{\mathbf{F}}^{h,e}$.

5.4.9. Stability

The last but not least property of the novel finite element we discuss is its stability. In particular we are concerned with *hourglassing-instabilities* from which almost all EAS elements suffer. To the best knowledge of the authors the only exemption is the element for explicit dynamics proposed in the PhD-Thesis of Schmied [125] which is based on h-enhancement similarly to H1U/h12.

The numerical investigations in Section 5.5 (especially Sections 5.5.6 to 5.5.9) cover many cases in which other EAS elements and mixed approaches famously exhibit hourglassing instabilities. The newly proposed element H1U/h12 passes all these tests without any problems which strongly suggests its improved stability. Yet, it is of course no mathematically sound proof which is beyond the scope of the present contribution. However, we suggest:

Hypothesis 5.1. *Satisfaction of the piecewise constant pressure orthogonality conditions* (5.64) *is a necessary condition for stability.*

We back our claim by recalling that in the very first publication on geometrically linear EAS elements Simo and Rifai [134] already observed spurious oscillations of the variationally consistent pressure field. Later, Reddy and Simo [120] provided a mathematical proof revealing that the independent stress field of EAS elements exhibits *checkerboard modes* similar to the Q1/P0 element. These instable modes exist due to a violation of the inf-sup condition which is thoroughly covered by Boffi et al. [23] and for EAS elements in [74, 120]. Fortunately, in case of the standard linear elastic EAS element [134], the checkerboard modes are confined to the independent stress. The displacement as well as the enhanced strain are unaffected (see Lamichane et al. [74] and Djoko et al. [39]). Thus, by either employing L_2-smoothing of the independent stress [134] or by using the constitutive stress for post-processed results, the instable checkerboard modes do not affect linear EAS elements in practical simulations.

Returning now to nonlinear EAS elements we observe from (5.65) that the enhanced field is coupled with piecewise constant pressure (checkerboard modes) in case of F-enhancement. Our suggestion is that this leads to instabilities from the independent stress field being transferred to the enhanced field and ultimately the well-known hourglassing first observed by Wriggers and Reese [156]. The coupling between enhanced fields and checkerboard modes can be avoided with h-enhancement since it is simple to satisfy (5.64). In particular, H1U/h12 fulfills (5.64) at least for regular meshes. Thus, the enhanced field is decoupled from *independent* constant pressure and instabilities *cannot* be transmitted from the stress to the strain field. We believe that this explains the *increased stability* of H1U/h12.

Remark 5.5. *Unfortunately, satisfaction of Hypothesis 5.1 is not sufficient. E.g., if* $\mathbf{F}^{h,e}$ *is used instead of* $\mathbf{F}_0$ *in* (5.22) *we observe instabilities (only) for* rectangular *element shapes in the one element stability analysis (see Section 5.5.6) even though the element still satisfies* (5.64).

5.5. Numerical investigations

This section covers extensive numerical studies to asses the performance of the newly proposed element **H1U/h12** (Q1U/h4)[11] and compare it to existing elements. The numerical examples cover a wide range of features including mesh distortion sensitivity, robustness and stability in hyperelastic as well as elasto-plastic simulations. Focus is put on three-dimensional hexahedral elements. However, results can at least qualitatively be transferred to two-dimensional plane-strain problems if not mentioned otherwise.

Element types

The (standard) Bubnov-Galerkin finite elements used in our numerical investigations for comparison with H1U/h12 are:[11]

- **H1/E9** (Q1/E4): Standard EAS element by Simo and Armero [131] with the classical nine (four in 2D) Wilson-modes and F-enhancement.
- **HA1/E12T** (QA1/E4T): Standard EAS element using the transposed Wilson-modes to avoid the instability under compression and three additional volumetric enhanced modes. Additionally, a modification of the gradient of the compatible shape functions and a special nine point (five in 2D) Gauss quadrature rule are used. The

[11] Element names in parentheses denote corresponding 2D elements.

element is implemented as described by Pfefferkorn and Betsch [104] and closely related to the improved EAS version by Simo et al. [132].

- **H1/P0** (Q1/P0): Mixed pressure element by Simo et al. [135] based on a Hu-Washizu functional with elementwise constant pressure field.
- **H1/P0E6T** (Q1/P0E2T): Combination of H1/P0 and H1/E9T which was proposed by Armero [8] for 2D plane strain problems (Q1/P0E2T). A 3D extension (H1/P0E6T) has recently been proposed by Hille et al. [52].
- **H1/S18** (Q1/S5): Assumed stress element as proposed by Pian and Sumihara [110] and Pian and Tong [112] for linear elasticity in 2D and 3D, respectively. An extension to nonlinear problems can, e.g., be found in the work of Viebahn et al. [145]. However, we use the inverse stress strain relation for a Neo-Hookean material model proposed by Pfefferkorn et al. [109] instead of the numeric procedure in aforementioned reference.

We also consider other *Petrov-Galerkin enhanced assumed strain (PG-EAS)* elements in addition to H1U/h12. All of which use the same set of ansatz functions (see Section 5.3.2) and differ only in the type of enhancement and required transformations for objectivity. In particular we investigate:

- **H1U/E12** (Q1U/E4): E-enhancement (see Section 5.3.3.3).
- **H1U/F12** (Q1U/F4) and **H1U/F12T** (Q1U/E4T): F-enhancement as described in Section 5.2.3, Remark 5.3 and Section 5.3.3.2. H1U/F12T uses the transposed Wilson-modes (i.e., the transpose of (5.33)) only for the trial function of the enhancement.

Material models

For the *hyperelastic* simulations we consider a Neo-Hookean (NH) and Ogden (OG) material. The former is chosen as standard material model and the latter allows to deliberately construct a material with an instability under tension. Such instabilities induce hourglassing for some elements which is thoroughly covered in Section 5.5.6 and 5.5.7. The strain-energy functions of the two hyperelastic material models are given by

$$W_{\mathrm{NH}} = \frac{\mu}{2}\left(\mathrm{tr}(\mathbf{F}^{\mathrm{T}}\mathbf{F}) - 3\right) + \frac{\lambda}{2}\ln^2(J) - \mu\ln(J), \tag{5.66a}$$

$$W_{\mathrm{OG}} = \sum_{j=1}^{n_p}\sum_{i=1}^{3}\frac{\mu_p}{\alpha_p}\left(\bar{\lambda}_i^{\alpha_p} - 1\right) + \kappa\beta^{-2}\left(\beta\ln(J) + J^{-\beta} - 1\right), \tag{5.66b}$$

where $J = \det \mathbf{F}$ is the determinant of the deformation gradient (5.39) and $\bar{\lambda}_i = J^{-\frac{1}{3}}\lambda_i$, $i = 1, 2, 3$ are the deviatoric parts of the principal stretches λ_i. Furthermore, λ, μ and κ are the two Lamé-constants and the bulk modulus and β, α_p, μ_p, n_p are further material constants specified in the respective examples. The eigenvalue decomposition required for the Ogden material is carried out using a numerical eigenvalue and eigenvector computation and proper treatment of duplicate eigenvalues for the material tangent (see, e.g., [25, 53]).

In addition to the hyperelastic material, we consider the eigenvalue based *elasto-plastic* (EP) material model proposed by Simo [130]. It contains the multiplicative elasto-plastic split, nonlinear isotropic hardening, the von Mises yield condition and a Hencky elastic law. The same eigenvalue routine as described for the Ogden model is used. This material is a standard model in finite element technology (e.g., [8, 65, 105, 132]) and employed with the standard material parameters $\mu = 80.1938$, $\kappa = 164.206$, $\sigma_{Y0} = 0.45$, $\sigma_{Y\infty} = 0.715$, $\delta = 16.93$ and $H = 0.12924$ (e.g., [8, 47, 67, 105, 129–132, 134]). For all elasto-plastic simulations we use the *line-search* algorithm described by Bonet and Wood [25] to stabilize the Newton-Raphson scheme. Hyperelastic simulations do *not* employ the line-search algorithm.

Table 5.1 gives an overview of the material models used for the numerical examples in this section.

Table 5.1.: Overview of the material models employed for the various benchmarks. Neo-Hookean (NH), Ogden (OG) and elasto-plastic (EP) material model.

benchmark	section	material
patch test	5.5.1.1	NH, EP
frame invariance	5.5.1.2	NH
objectivity	5.5.1.3	NH, EP
eigenvalues	5.5.1.4	NH
mesh distortion	5.5.2	NH
roll-up	5.5.3	NH

benchmark	section	material
Cook's membrane	5.5.4	NH
thin circular ring	5.5.5	NH
one element stability	5.5.6	OG
large mesh stability	5.5.7	OG
necking plane strain	5.5.8	EP
necking circular bar	5.5.9	EP

Stress recovery

All stress results shown in the present work are computed using a L_2-smoothing procedure to project the stress from the Gauss points to the nodes. It is given by

$$\sum_{e=1}^{n_{\mathrm{el}}} \tau_i \int_{\Omega^e} N_i N_j \,\mathrm{d}V \sigma_j = \sum_{e=1}^{n_{\mathrm{el}}} \tau_i \int_{\Omega^e} N_i \hat{\sigma} \,\mathrm{d}V, \qquad \forall\ \tau_i \in \mathbb{R} \tag{5.67}$$

where N_i are the standard Lagrangian shape functions, σ_j the stress at node j and $\hat{\sigma}$ the constitutive stress[12]. Usually, we plot results for the von Mises stress. However, (5.67) can be used for any stress component.

5.5.1. Basic tests

In this section we briefly summarize results of some fundamental numerical tests for H1U/h12. Full descriptions of the tests' setup and thresholds to verify if the test is passed can be found in the references mentioned below.

[12] For H1/S18 we use the independent stress field.

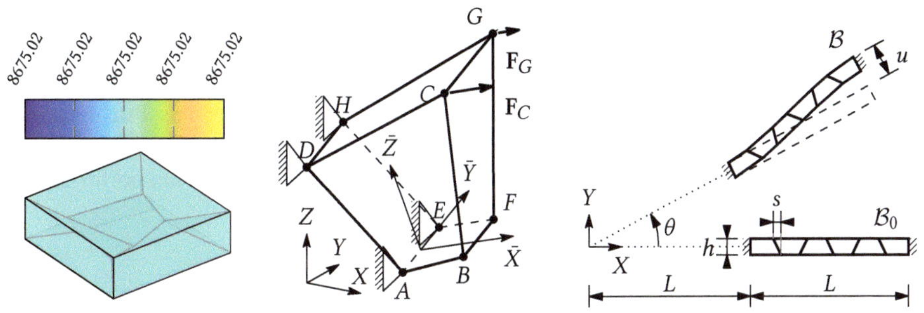

Figure 5.2.: Von Mises stress distribution for the patch test with distorted mesh (left). Geometry and boundary conditions for the isotropy and invariance test (middle) and the objectivity test (right).

5.5.1.1. Patch test

We conduct the numerical patch test exactly as described by Pfefferkorn and Betsch [104–106]. Element H1U/h12 passes the test for elastic and elasto-plastic material as expected from the analytic investigation in Section 5.4.6. See Figure 5.2 for a stress plot computed with the novel finite element.

5.5.1.2. Frame invariance and isotropy

H1U/h12 also passes the frame invariance and isotropy benchmark which is fully described in Pfefferkorn and Betsch [106]. Figure 5.2 shows the single finite element used to check the two properties. Corresponding analytic proofs that the element passes this numerical test are presented in Sections 5.4.2 and 5.4.3.

5.5.1.3. Objectivity

Another basic test concerns the objectivity of H1U/h12 which is investigated by examining the effect of superimposed rigid body motions on a beam-like structure (see Figure 5.2). A full description of the test is given in [105]. The only difference in this work is the use of distorted elements ($s = h/2$) as shown in Figure 5.2 to ensure that the metric ansatz functions do not coincide with the standard ansatz functions. Furthermore, as described by Pfefferkorn and Betsch [105], the test also covers path independence. Again, element H1U/h12 passes both tests for elastic and plastic materials which is inline with Sections 5.4.2 and 5.4.3.

5.5.1.4. Eigenvalue analysis

A final simple numerical benchmark concerns the eigenvalue analysis of a single (regular or distorted) finite element in the reference configuration which can be used to determine whether or not the element is prone to volumetric locking. Refer to Pfefferkorn and Betsch [104–106] for a full description of this benchmark. In this test H1U/h12 exhibits the correct distribution of eigenvalues with only one "locking" mode regardless of mesh distortion. However, as analogously discovered for the corresponding linear elastic element [106], H1U/h12 exhibits two eigenvalue pairs with non-negligible complex part. Fortunately, this does not seem to have negative effects on the elements performance as our extensive numerical studies show.

5.5.2. Mesh distortion test

The first benchmark for features beyond basic requirements concerns mesh distortion sensitivity. To that end we consider a 3D version [104, 106] of the standard 2D test [3, 6, 67, 70, 113, 115, 134]. The beam like structure shown in Figure 5.3 has the dimensions $l \times b \times h = 10 \times 1 \times 2$ and is meshed with only two elements. Distortion is applied on two nodes via parameter s as shown in Figure 5.3. Fixed boundary conditions $u(X=0, Y, Z) = 0$, $v(X=0, Y=0, Z) = 0$ and $w(X=0, Y, Z=0) = 0$ apply on the left end ($X = 0$) while the other end is subjected to a bending moment $M = 20$ applied in the form of a traction boundary condition. It is given by $\sigma(Z) = 30\,(1-Z)$ and modeled as dead load. We choose the Neo-Hookean material model with parameters $\mu = 600$ and $\lambda = 600$ for this test which corresponds to $E = 1500$ and $\nu = 0.25$.

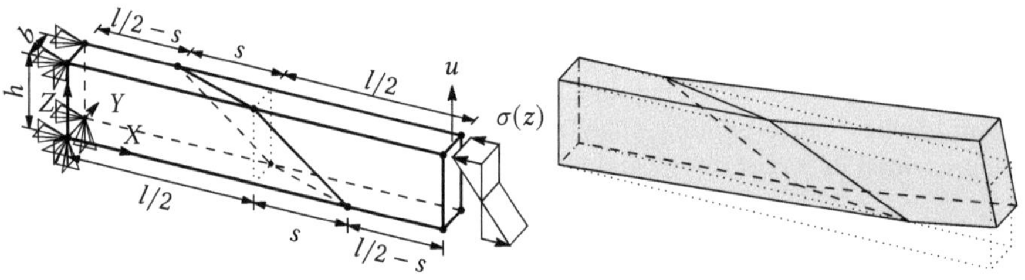

Figure 5.3.: Mesh distortion test. Setup (left) and deformed configuration for $s = 3$ (right). Deformed configuration computed with H1U/h12.

The setup is chosen such, that the analytic solution in linear elasticity is $u = 1$ (cf. Jabareen and Rubin [62]). However, for the nonlinear problem we use $u_{\text{ref}} = 0.96897$ computed with a fine regular mesh of $40 \times 8 \times 4$ HA1/E12 elements as reference solution. Displacement u denotes the mean value of the displacement in Z-direction of the two nodes at the top edge on the right face (see Figure 5.3) which is normalized with u_{ref}. Figure 5.4 shows the resulting $\delta = u/u_{\text{ref}}$ for several elements.

Element H1/S18 exhibits the best performance of the standard elements with almost the correct displacement ($|\delta - 1| < 1\%$) for no distortion and the least deterioration with mesh distortion. H1/P0E6T behaves far too soft in case of no distortion. In fact there is a "optimal" distortion for which the element is exact, which explains the great performance of that element in the Cook's membrane example in Section 5.5.4.

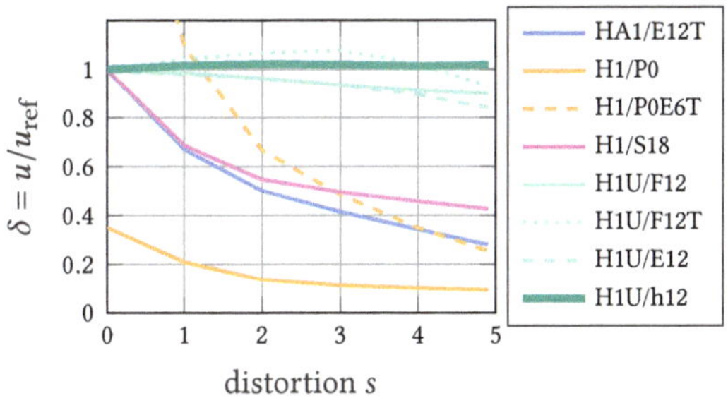

Figure 5.4.: Mesh distortion test. Normalized top edge displacement δ plotted over distortion s.

The novel PG-EAS elements perform far better than any of the standard elements. After all, they are specifically crafted to be exact in linear elastic bending problems (see Section 5.4.5) which also greatly improves their accuracy in this nonlinear test. However, results are not as accurate as in the linear elastic case and there are, as expected (see Section 5.4.5.2), small deviations from the optimal result $\delta = 1$. Of the tested PG-EAS elements the newly proposed H1U/h12 is the most accurate with $|\delta - 1| < 2.1\%$ for all distortions. The other PG-EAS elements are slightly less accurate but still outperform all standard Bubnov-Galerkin finite elements.

Remark 5.6. *Interestingly, none of the PG-EAS elements converges in the Newton-Raphson scheme regardless of number of load steps for the two special meshes proposed by Pfefferkorn and Betsch [106]. There seems to be some sort of instability which might be associated with horizontal displacement observed due to the combination of non-symmetric Dirichlet boundary conditions and mesh distortion. If either a regular or finer mesh is considered, the solution converges as usual.*

5.5.3. Roll-up

Our next example concerns the classical roll-up of a beam to show that the novel approach also improves the element's behavior in distorted meshes for considerable bending. Similar to the previous example we study a beam-like structure with dimensions $L = 10$, $b = 2$ and $t = 0.5$ which is subjected to a bending moment

$$M = \frac{2\pi EI}{L}\left(1 - \frac{2}{3}\frac{\pi^2 t^2}{L^2}\right), \qquad I = \frac{bt^3}{12} \tag{5.68}$$

applied on the right end in the form of a linearly distributed follower load. Thus, it stays normal to the deformed surface and is scaled with the deformed area (see, e.g., Wriggers [154] Ch. 4.2.5). The moment (5.68) is taken from the work of Müller and Bischoff [94] and accounts for the rather thick beam in contrast to the usually used $M = 2\pi EI/L$ which is only valid for thin beams. Dirichlet boundary conditions $u(X = 0, Y, Z) = 0$, $v(X = 0, Y = 0, Z) = 0$ and $w(X = 0, Y, Z = 0) = 0$ apply on the left end and we consider a regular and distorted mesh with 20 elements as shown in Figure 5.5. The material parameters of the Neo-Hookean model are chosen to $\mu = 500$ and $\lambda = 0$ which correspond to $E = 1000$ and $\nu = 0$.

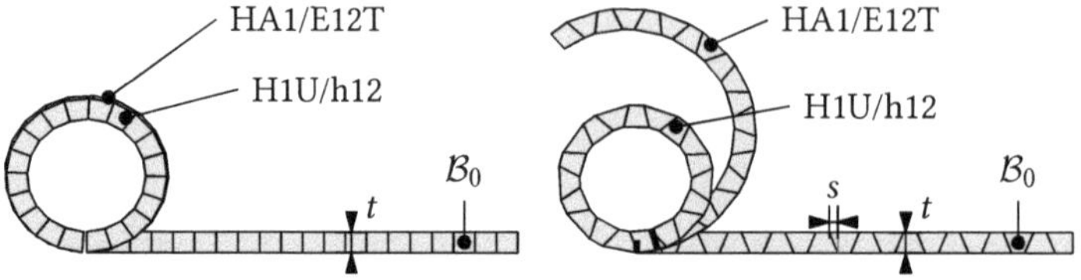

Figure 5.5.: Roll-up of a beam. Reference $\mathcal{B}_0$ and deformed configuration for HA1/E12T and H1U/h12 for a regular (left) and distorted (right) mesh. Thick dashed and solid line show beginning and end of the beam with H1U/h12 and the distorted mesh (right).

Figure 5.5 shows the final configuration computed with HA1/E12T and H1U/h12. The former is only capable of (almost) correctly depicting the roll-up into a circular shape in case of a regular mesh and is well off the desired shape in case of distortion. However, the newly proposed H1U/h12 is able to give good results for both meshes with only slightly too soft behavior due to the distortion. Thus, it can be concluded, that the Petrov-Galerkin approach improves performance not only for "moderate" bending (see Section 5.5.2) but also in case of much larger curvatures.

5.5.4. Cook's membrane

The classical Cook's membrane example (e.g., [8, 47, 63, 64, 104–106, 131, 134, 141]) covers convergence of displacement and stress with mesh refinement as well as coarse mesh accuracy. Figure 5.6 shows the tapered trapezoidal specimen which is clamped on the left side ($\mathbf{u}(X=0, Y, Z) = \mathbf{0}$) and subjected to a constant shear force $\tau = 100$ in y-direction. Again, we use the Neo-Hookean material with the parameters chosen to $\lambda = 8.2669 \cdot 10^4$ and $\mu = 756.00$ (corresponding to a nearly incompressible material with $E = 2261.2$ and $\nu = 0.4955$). A mesh with two elements in direction of thickness and $n_{el} = \{2, 4, 8, 16\}$ elements in the other two directions completes the setup.

Figure 5.7 shows the displacement u of the top right corner in Y-direction and the stress σ at the midpoint of the lower surface (see Figure 5.6) for increasingly fine meshes. H1/P0E6T exhibits seemingly the best results of the standard elements. However, as shown in Section 5.5.2 it is too soft in bending dominated problems with moderate distortion. Thus, the good results of H1/P0E6T in this test are due to the "correct distortion" of the mesh. H1/S18 is the next best standard element and shows especially fast stress convergence which even competes with the results of the Petrov-Galerkin finite elements.
All unsymmetric elements exhibit fairly similar results with an odd kink for very coarse meshes. The novel H1U/h12 exhibits fast convergence of the displacement and only marginally slower convergence of the stress in comparison to H1/S18. All in all, the newly proposed element gives accurate results in this test as well.

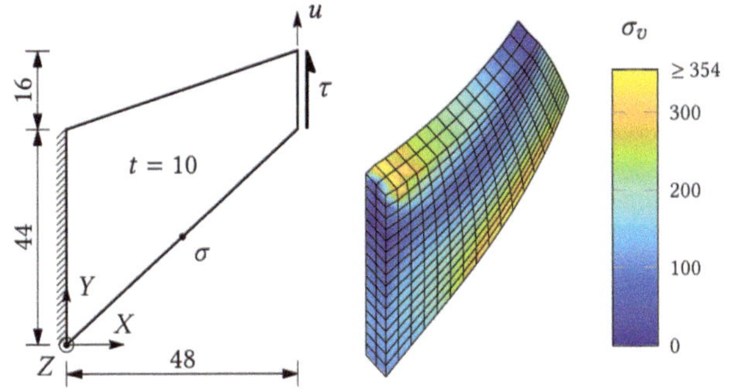

Figure 5.6.: Cook's membrane test. Setup (left) and deformed configuration with von Mises stress distribution computed with 16 × 16 × 2 H1U/h12 elements (right). The colors are capped at $0.65\sigma_{v,\max} = 354$ to avoid domination of the stress singularity in the plot.

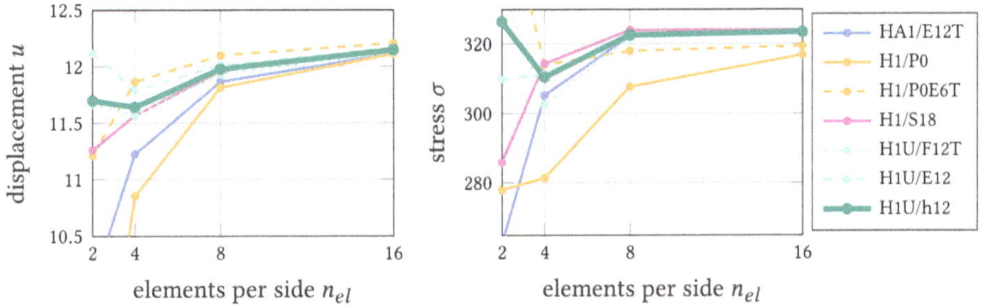

Figure 5.7.: Cook's membrane test. Convergence of the displacement u (left) and stress σ (right) with mesh refinement (h-convergence).

5.5.5. Thin circular ring

In this Section we conduct the *thin circular ring* example [13, 67, 109] shown in Figure 5.8 to examine the element's robustness in the Newton-Raphson scheme. We characterize robustness by the size of applicable load steps and number of Newton-Raphson iterations required for convergence [109]. Robustness is naturally highly influenced by the settings for Newton-Raphson scheme. We choose a convergence criterion based on the norm of the residual $||\mathbf{R}|| < 10^{-8}$ and assert failure of the Newton-Raphson scheme if either $||\mathbf{R}|| > 10^{14}$ or more than 20 iterations are necessary for convergence within one load step. Figure 5.8 shows the slit circular ring with dimensions $r_i = 6$, $r_o = 10$ and $t = 0.03$. Boundary conditions $u(0,0,0) = 0$, $v(X,Z,Y{=}0) = 0$ and $w(X,Y{=}0,Z{=}0) = 0$ apply at face F_1 and face F_2 is subjected to a dead load $q = 6.67$ in Z-direction. To complete the setup, we consider the Neo-Hookean material model with $\mu = 10.5 \cdot 10^3$ and $\lambda = 0$.

The converged displacement w_P of point P in Z-direction computed with a very fine mesh with $8 \times 24 \times 120$ elements is $w_P \approx 10.265$ for both H1/S18 and H1U/h12. However, Figure 5.8 shows the deformed configuration for the coarser mesh with $2 \times 6 \times 30$ elements

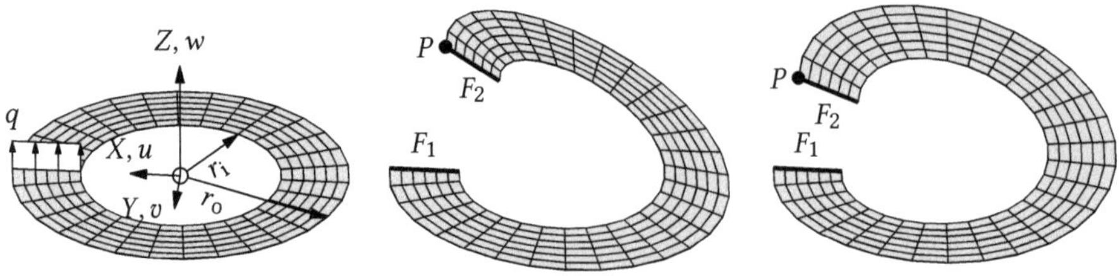

Figure 5.8.: Thin circular ring. Geometry (left) and deformed configuration computed with $2 \times 6 \times 30$ H1/S18 (middle) and H1U/h12 (right) elements.

for H1/S18 with $w_P = 9.741$ and H1U/h12 with $w_P = 7.978$, respectively. Therefore, both behave too stiff and the different displacements make comparison regarding robustness difficult. In order to get a "fair" comparison of the element's robustness we only consider a mesh with $4 \times 12 \times 60$ elements for which the displacements w_P (see Table 5.2) are similar.

Table 5.2 also lists the required number of load steps n_{steps} and total number of Newton-Raphson iterations n_{NR} for various element types. H1/S18 exhibits the most favorable behavior with only one required load step and seven Newton-Raphson iterations. This is inline with the results of Magisano et al. [87] and Pfefferkorn et al. [109] who observed the high robustness of assumed stress elements in many examples. All EAS elements require many more load steps and iterations. However, this can be greatly improved with the MIP method briefly described in Section 5.3.4 and introduced for EAS elements by Pfefferkorn et al. [109]. For the novel H1U/h12 the MIP method allows to reduce n_{NR} by a factor of ≈ 5.6.

Table 5.2.: Results of the thin circular ring test.

element type	req. n_{steps}	total n_{NR}	w_P
H1/S18	1	7	10.235
HA1/E12T	5	80	9.970
HA1/E12T-MIP	2	14	9.970
H1U/E12	8	106	10.286
H1U/F12	8	107	10.292
H1U/F12T	7	96	10.224
H1U/h12	6	90	10.200
H1U/h12-MIP	2	16	10.200

5.5.6. One element stability analysis

The next two tests concern the stability of the finite elements and are used to determine if elements are prone to hourglassing. To that end we first consider the unconstrained one element stability test proposed by Glaser and Armero [8] (see also Armero [8]) which has recently been extended to 3D problems by Pfefferkorn and Betsch [105]. The advantage of the unconstrained compared to the constrained test proposed by Wriggers and Reese [156] is that it can also be used to examine hourglassing due to material instabilities.

We consider the Ogden material model (5.66b) with $\beta = 2$, $\kappa = 10^5 + 20/3$, $n_p = 1$, $\alpha_1 = 0.5$, $\mu_1 = 80$ which corresponds to $\mu = 20$ and $\lambda = 10^5$ in linear theory. Note that the material parameters are deliberately chosen such that a material instability under tension occurs at $\lambda_1^{\text{crit}} \approx 3$ and $\lambda_1^{\text{crit}} \approx 1.75$ for the first and second Piola-Kirchhoff stress, respectively (cf. Glaser and Armero [47]). This instability is similar to the one of the elasto-plastic material model which is well-known [8, 47, 67, 105] to trigger hourglassing. However, the Ogden model yields more distinct results which is why it is employed here.
The test is performed on a single element (see Figure 5.9) with reference configuration $\Omega_0^e = [-r, +r] \times [-1, +1] \times [-1, +1]$ where r governs the element's aspect ratio. Usually, it is not necessary to consider initially rectangular geometries with $r \neq 1$ since Armero [8] showed that different aspect ratios only change when[13] and not if instabilities occur. However, this is not necessarily true for some of the unsymmetric finite elements which is discussed below.

The homogeneous deformation state shown in Figure 5.9 is associated with a diagonal deformation gradient of the form $\mathbf{F} = \text{diag}([\lambda_1, \lambda_2, \lambda_3])$. Therein, λ_i, $i = 1, 2, 3$ are the principal stretches associated with the principal axes which coincide with coordinate system shown in Figure 5.9. Furthermore, choosing a specific λ_1 allows to compute the other principal stretches from the material model and the boundary conditions. In case of a 2D plane strain problem we have $\lambda_3 = 1$ and λ_2 can be computed from $\tau_2 = 0$. Similarly, in 3D uniaxial tension $\lambda_2 = \lambda_3$ can be determined from $\tau_2 = \tau_3 = 0$.

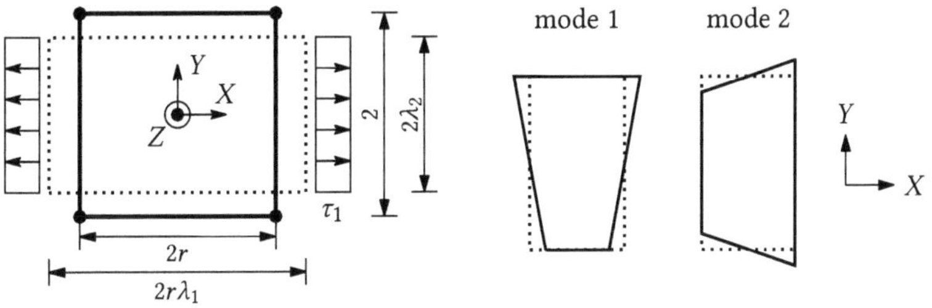

Figure 5.9.: One element stability test. Geometry (left) and 2D hourglass modes computed for $\lambda_1 = 0.75$ (right, solid line). The deformed state is depicted in both figures with dotted lines.

[13] See also Sussmann and Bathe [138] who show that instabilities can even occur for small deformations if the element's aspect ratio is high.

This analytic solution can then be imposed on the finite element in order to compute the corresponding stiffness matrix $\mathbf{K}^e$ of the element. Furthermore, the hourglass eigenvectors of $\mathbf{K}^e$ arranged row-wise in matrix $\mathbb{P}_{\text{hour}}$ can be determined in closed form for this example. In 2D plane strain they are always given by

$$\mathbb{P}_{\text{hour}} = \frac{1}{4}\begin{bmatrix} 1 & 0 & -1 & 0 & 1 & 0 & -1 & 0 \\ 0 & 1 & 0 & -1 & 0 & 1 & 0 & -1 \end{bmatrix}. \tag{5.69}$$

Unfortunately, no equivalently simple structure exists for the 3D problem. Nevertheless, it is possible to find the eigenvectors in closed form with the procedure described by Pfefferkorn and Betsch [105] for symmetric elements. Interestingly, the approach can also be used for most of PG-EAS elements since their stiffness matrix is symmetric for the present load state and regular mesh. However, in case of H1U/h12 $\mathbf{K}^e$ looses its symmetry and the method has to be modified. The eigenvectors are then given by [105]

$$\mathbb{P}_{\text{hour}} = \begin{bmatrix} 0 & 0 & 0 & 0 & 0 & 0 & 0 & 0 & 0 & 1 & 0 & 0 \\ 0 & 0 & 0 & 0 & 0 & 0 & 0 & 0 & 0 & 0 & 1 & 0 \\ 0 & 0 & 0 & 0 & 0 & 0 & 0 & 0 & 0 & 0 & 0 & 1 \\ 0 & 0 & 0 & 0 & -1 & 0 & 0 & 0 & 1 & 0 & 0 & 0 \\ 0 & 0 & 0 & 0 & 0 & -1 & 0 & 1 & 0 & 0 & 0 & 0 \\ 0 & 0 & 0 & 0 & 0 & 1 & 0 & 1 & 0 & 0 & 0 & 0 \\ 0 & U_1 & 0 & 1 & 0 & 0 & 0 & 0 & 0 & 0 & 0 & 0 \\ 0 & U_2 & 0 & 1 & 0 & 0 & 0 & 0 & 0 & 0 & 0 & 0 \\ 0 & 0 & U_1 & 0 & 0 & 0 & 1 & 0 & 0 & 0 & 0 & 0 \\ 0 & 0 & U_2 & 0 & 0 & 0 & 1 & 0 & 0 & 0 & 0 & 0 \\ V_1 & 0 & 0 & 0 & 1 & 0 & 0 & 0 & 1 & 0 & 0 & 0 \\ V_2 & 0 & 0 & 0 & 1 & 0 & 0 & 0 & 1 & 0 & 0 & 0 \end{bmatrix} \underbrace{\frac{1}{8}\begin{bmatrix} \mathbf{h}_1 \odot \mathbf{I} \\ \vdots \\ \mathbf{h}_4 \odot \mathbf{I} \end{bmatrix}}_{\mathbb{H}_{\text{hour}}}, \tag{5.70}$$

where $\mathbf{I}$ is the identity matrix, $\mathbf{h}_i$ are the hourglass vectors given in (5.20c) and "$\odot$" denotes the Kronecker product [79]. Furthermore, U_i and V_i, $i = 1, 2$ can directly be computed from the components of the sparse [105] matrix $\mathbf{k}^e = [k_{ij}] = \mathbb{H}_{\text{hour}}\mathbf{K}^e\mathbb{H}_{\text{hour}}^{\text{T}}$ and are given by

$$U_{1,2} = -\frac{p_1}{2} \pm \sqrt{\frac{p_1^2}{4} + 1}, \qquad p_1 = \frac{k_{44} - k_{22}}{k_{24}}, \tag{5.71a}$$

$$V_{1,2} = -\frac{p_2}{2} \pm \sqrt{\frac{p_2^2}{4} + 2\frac{k_{15}}{k_{51}}}, \qquad p_2 = \frac{k_{55} - k_{11} + k_{59}}{k_{51}}. \tag{5.71b}$$

Thus, the only difference to the method proposed by Pfefferkorn and Betsch [105] is that it considers the *only* unsymmetric entry $k_{15} \neq k_{51}$ of $\mathbf{k}^e$. Figure 5.9 and 5.10 show the computed eigenmodes for $\lambda_1 = 0.75$. Ultimately, the hourglass eigenvectors can be used to evaluate

$$\text{diag}([\omega_i^{\text{hour}}]) = (\mathbb{P}^{\text{hour}})^{-\text{T}}\mathbf{K}^e(\mathbb{P}^{\text{hour}})^{\text{T}} \tag{5.72}$$

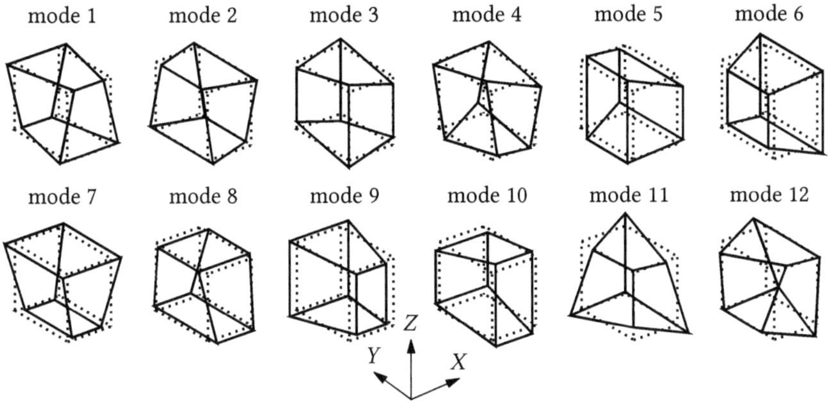

Figure 5.10.: One element stability test. Eigenmodes in 3D with non-constant strain field (solid line) of H1U/h12 for the deformation state $\lambda_1 = 0.75$ (dotted). Computed using the Ogden material model.

which yields a diagonal matrix[14] with the eigenvalues ω_i^{hour}. If any negative hourglass eigenvalues occur the element is likely to be prone to hourglassing. Whether or not hourglass modes actually occur can then be verified with the simulation described in Section 5.5.7.

Figure 5.11 shows the hourglass eigenvalues ω_i^{hour} for several 2D elements in dependence of λ_1. The F-enhanced PG-EAS Q1U/F4 and Q1U/E4 both coincide with Q1/S5 and are therefore not included in the plots. These three elements all exhibit the well-known instability of ω_2^{hour} (solid line) under compression ($\lambda_1 < 1$ implies compression) first discovered for EAS elements by Wriggers and Reese [156] (see also Viebahn et al. [145] for assumed stress elements). All other elements exhibit no negative eigenvalues under compression and are therefore not prone to hourglassing.
In case of tension ($\lambda_1 > 1$) instabilities can be observed due to the material instability. In fact, the only EAS element without zero eigenvalue is the newly proposed Q1U/h4. Interestingly, the zeros of the other elements occur at $\lambda_1 \approx 1.75$ and $\lambda_1 \approx 3$ which coincides with instability points of the Ogden material. This supports the claim that the material instabilities transfer to the hourglass modes and cause the hourglassing patterns (cf. Glaser and Armero [47]). Hourglass mode ω_1^{hour} (dashed line) also becomes negative for some finite elements under tension but not for Q1U/h4.
The second plot of Figure 5.11 shows the hourglassing eigenvalues of Q1U/h4 with various element aspect ratios r. As mentioned above, some unsymmetric finite elements exhibit instabilities for rectangular elements even if they are stable for the quadratic element. This is, e.g., the case if $\mathbf{F}^{\text{h},e}$ is used instead of $\mathbf{F}_0$ in (5.22). Fortunately, Q1U/h4 does not suffer from such defects.

[14] If the matrix is truly diagonal can be used to check if the test is performed correctly and verify that $\mathbb{P}^{\text{hour}}$ are actually eigenvectors of $\mathbf{K}^e$.

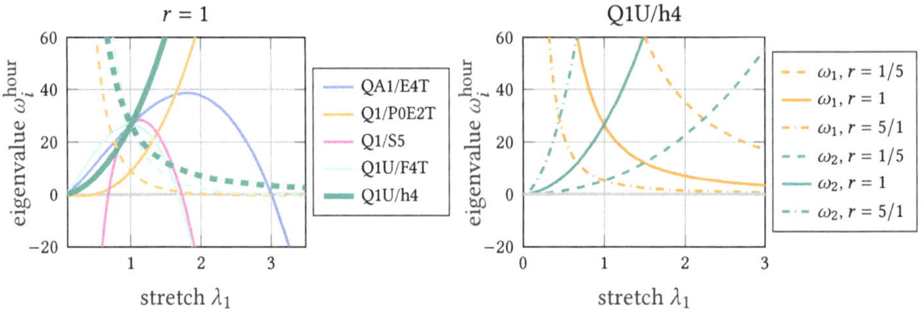

Figure 5.11.: One element stability test. Results of the 2D analysis. Left: Eigenvalues ω_1^{hour} (dashed) and ω_2^{hour} (solid) for various elements with $r = 1$. Right: Eigenvalues ω_i^{hour} of Q1U/h4 for various aspect ratios r.

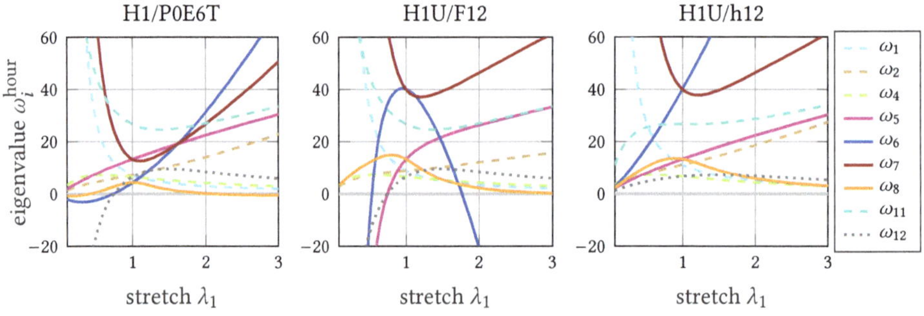

Figure 5.12.: One element stability test. Results of the 3D analysis for $r = 1$. Hourglass eigenvalues for the Ogden material and H1/P0E6T (left), H1U/F12 (middle) and H1U/h12 (right).

Figure 5.12 shows results of the 3D one element stability test for the elements H1/P06T, H1U/F12 and H1U/h12. The eigenmodes 3, 9 and 10 are not included in the plots since they are duplicates[15] of 2, 7 and 8, respectively. Almost all elements (not only the plotted) exhibit an instability for mode 12. The only exemption is the newly proposed H1U/h12 which is an interesting result even if mode 12 does usually not lead to global hourglassing patterns since it is incompatible to neighboring elements [105]. The most important modes, 5-8, which are usually responsible for hourglassing, are highlighted with continuous lines in the plots. As expected, H1U/F12 exhibits instabilities both in compression (modes 5 and 6) and tension (mode 6). The instability of H1/P0E6T under compression in mode 6 does not seem to play any role in hourglassing problems. However, mode 8, which becomes "slightly" instable under tension, likely explains the hourglassing of H1/P0E6T observed in the elsto-plastic simulation shown in Section 5.5.9. Furthermore, the too soft bending behavior of H1/P0E6T (see Section 5.5.2) is resembled by the comparably soft behavior (low eigenvalue) of mode 7. The 3D results described for H1/P0E6T and H1U/F12 are inline

[15] Figure 5.10 indicates this since the corresponding modes are merely a rotation of each other around the x-axis.

with the observation by Hille et al. [52] for the elasto-plastic material. Both elements show hourglassing in such simulations. In contrast to that, the novel element H1U/h12 does not suffer from any instabilities in this test. While this is not enough to immediately conclude that H1U/h12 is completely stable (there could, e.g., be hourglassing in combination with neighboring elements), the simulations presented in the next sections further support the much improved stability of H1U/h12.

5.5.7. Large mesh stability analysis

In addition to the one element stability analysis in Section 5.5.6 we present a test on larger FE-meshes in this section. We focus on a 2D problem which is discussed in depth in Bieber et al. [20] and particularly interesting since there exists an *analytic* solution (see [98, 144]) of the considered diffuse bifurcation problem.

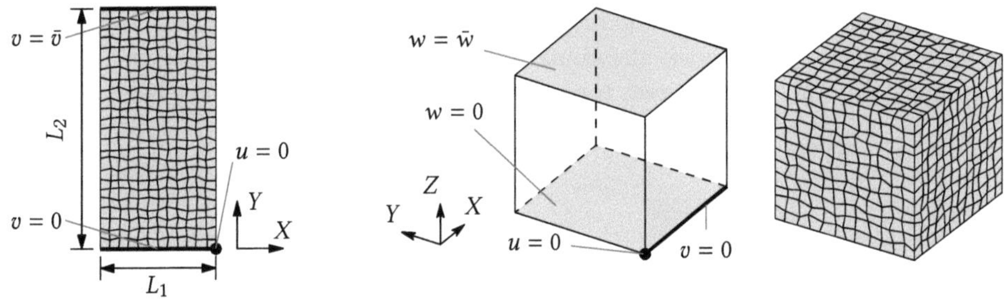

Figure 5.13.: Large mesh stability analysis. Setup and mesh of the large mesh stability analysis. Setup of 2D test with distorted mesh (left), setup for the 3D test (middle) and 3D distorted mesh (right).

The 2D version of the test is performed on the rectangular[16] block shown in Figure 5.13 with dimensions $L_1 = 25$ and $L_2 = 50$. The block is supported on the lower and top edge by the boundary conditions given in Figure 5.13 and is meshed with a regular or distorted mesh with 12×24 elements. The distortion is applied by randomly shifting all coordinates $|X_i - \bar{X}_i| \neq L_i/2$, where $\bar{X}_i$ is the block's center, by $\Delta_i \in [-0.5, 0.5]$ such that the surface is maintained. In order to also cover instabilities under tension the Ogden material (5.66b) is considered with the parameters $\beta = 2$, $\kappa = 3.333 \cdot 10^3$, $n_p = 1$, $\alpha_1 = 0.5$ and $\mu_1 = 1.379 \cdot 10^3$ (corresponding to $E = 1000$ and $\nu = 0.45$) which are chosen such that a material instability arises (see Section 5.5.6).

The actual stability analysis is performed by gradually de-/increasing the principal stretch in Y-direction by initially $\Delta\lambda = -0.01$ and $\Delta\lambda = +0.04$ in tension and compression, respectively. During that process the eigenvalues with lowest magnitude and the corresponding eigenvectors $\mathbf{\Phi}$ are computed numerically and tracked throughout the simulation which can be achieved by associating the current eigenvectors with the corresponding

[16] The rectangular shape makes it easier to compute the exact instability points since the (physical) modes do not all appear at almost the same load level (cf. Bieber et al. [20]).

eigenvectors from the previous load step. The step width $\Delta\lambda$ is automatically adapted to ensure that the correct eigenmodes are associated with each other. Whenever an eigenvalue changes its sign we start an *extended system* solver (e.g., Wriggers [154]). The especially taylored version used here solves

$$\begin{bmatrix} \boldsymbol{\varphi}^{\mathrm{h}} - \boldsymbol{\varphi}^{\mathrm{ana}}(\lambda) \\ \mathbf{K}_{\mathrm{red}}\boldsymbol{\Phi} + \alpha\boldsymbol{\Phi} \\ ||\boldsymbol{\Phi}||^2 - 1 \\ 0.5\alpha^2 \end{bmatrix} = \mathbf{0} \tag{5.73}$$

for the unknowns $\boldsymbol{\varphi}^{\mathrm{h}}$, $\boldsymbol{\Phi}$, α, λ which accurately determines the instability point with zero-eigenvalue and corresponding eigenmode $\boldsymbol{\Phi}$ as well as the critical stretch λ. In the equation above $\mathbf{K}_{\mathrm{red}}$ is the stiffness matrix where Dirichlet boundary conditions have been eliminated, $\boldsymbol{\varphi}^{\mathrm{ana}}$ the analytic solution of the homogeneous problem (see Section 5.5.6) and α is a regularization parameter which iterates to zero and is used to avoid ill-conditioning of the tangential matrix of (5.73). To compute the derivative of the stiffness matrix $\mathbf{K}$ required for the Newton-scheme we employ the numerical tangent procedure proposed by Wriggers and Simo [153]. The entire process is repeated until four instabilities have been found.

Q1/E4 **QA1/E4T**

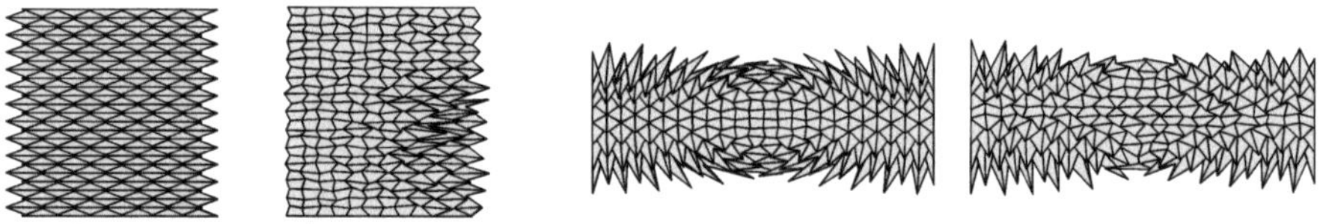

$\lambda_2^{\mathrm{regu}} = 0.7043$ $\lambda_2^{\mathrm{dist}} = 0.7107$ $\lambda_1^{\mathrm{regu}} = 2.8239$ $\lambda_1^{\mathrm{dist}} = 2.8252$

Figure 5.14.: Large mesh stability test. Results of the 2D analysis. Typical spurious modes for Q1/E4 under compression and QA1/E4T under tension (tension modes rotated by 90° and scaled in Y-direction by factor 4).

Figure 5.14 shows the first two modes of Q1/E4 under compression and QA1/E4T under tension, respectively. These elements exhibit typical spurious (hourglassing) modes which occur similarly for all standard and PG-EAS elements which do not employ h-enhancement. In particular, Q1/E4, Q1/S5, Q1U/E4 and Q1U/F4 exhibit instabilities under compression and tension while QA1/E4T, Q1/P0E2T and Q1U/F4T show no instabilities in the first four eigenmodes under compression but become unstable under tension due to the material instability. The only standard finite element that works well in this test is Q1/P0 which yields similar results as Q1U/h4 described below. These results are inline with Section 5.5.6.

The first two eigenmodes obtained with Q1U/h4 and corresponding analytically computed physical instabilities are shown in Figure 5.15. The analytic solution has first been proposed by Ogden [98] and is summarized well by Triantafyllidis et al. [144]. We also refer to Bieber et al. [20] for in-depth discussions and comparison of the analytical results to many standard finite element models. The numerical results of Q1U/h4 depicted in Figure 5.15

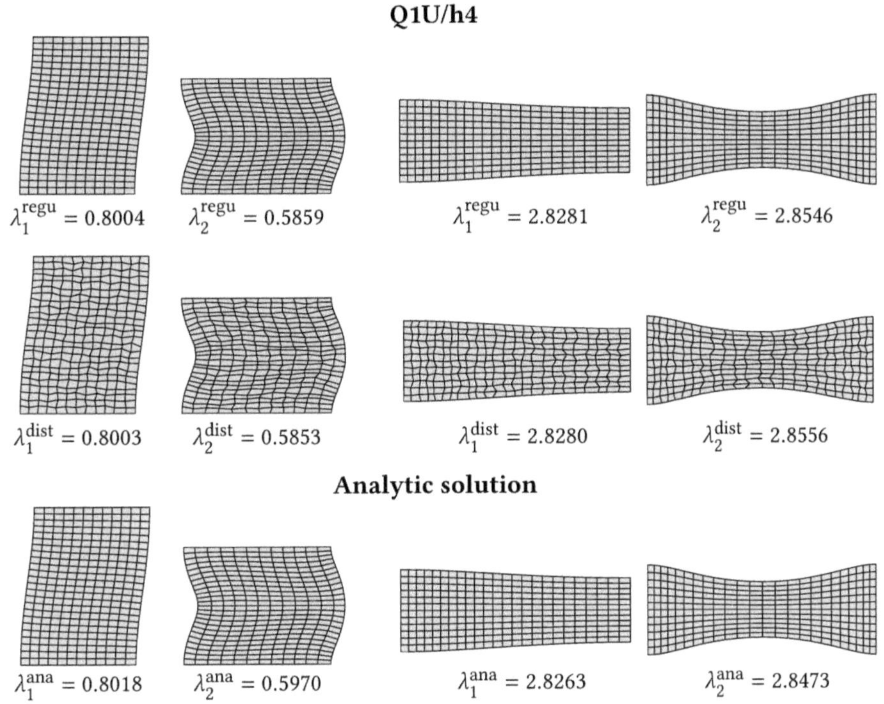

Figure 5.15.: Large mesh stability test. First two eigenmodes and corresponding critical stretches $\lambda_i^{(\bullet)}$ for the 2D large mesh stability analysis computed with Q1U/h4 for a regular (λ_i^{regu}, top row) and distorted mesh (λ_i^{dist}, middle row) are compared to the analytic solution (λ_i^{ana}, bottom row). All modes in tension are rotated by 90° and scaled in the Y-direction by factor 4.

clearly show that Q1U/h4 performs extremely well in this test. It reproduces the analytic results in terms of both the mode shapes and level of stretch at which the eigenmodes occur with high accuracy. The same holds also for the third mode (not shown) with the exemption of the third compression mode in combination with the distorted mesh. Only the fourth mode shape differs from the analytic result. Most importantly, no spurious modes have been observed for Q1U/h4 in all four modes.

For the 3D version of this test we study a cube with edge length of 50 and the boundary conditions shown in Figure 5.13. Similar to the 2D case we employ both a regular mesh with $12 \times 12 \times 12$ elements and a distorted mesh obtained by randomly shifting the nodes by $\Delta_i \in [-1, 1]$ (the same amount as in the 2D case in relation to the element size). The resulting eigenmodes computed with H1U/h12 are shown in Figure 5.16. Again, there are no spurious modes similar to the 2D case. Furthermore, the mode shapes as well as the critical stretches differ only slightly for the two meshes which highlights once more the mesh distortion insensitivity of the Petrov-Galerkin approach.

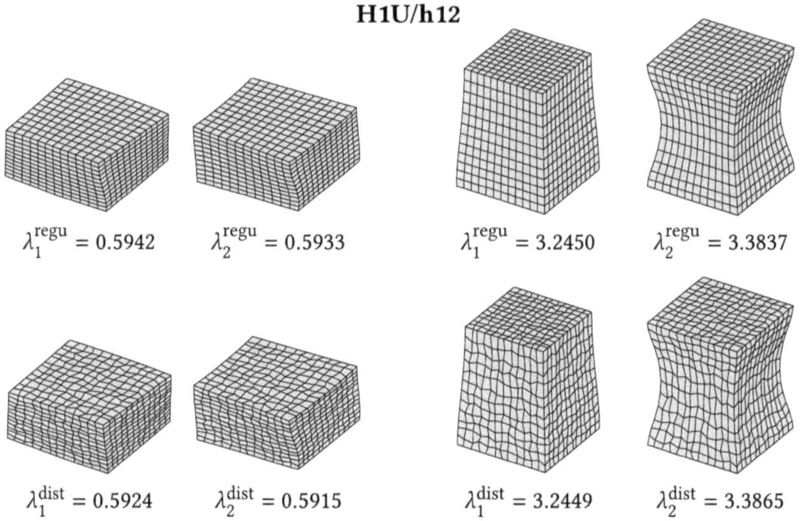

Figure 5.16.: Large mesh stability test. First two eigenmodes and corresponding critical stretches $\lambda_i^{(\bullet)}$ for the 3D large mesh stability analysis computed with H1U/h12 for a regular (λ_i^{regu}, top row) and distorted mesh (λ_i^{dist}, bottom row). Modes in tension are scaled in the Z-direction by factor 4.

5.5.8. Necking of a plane strain elasto-plastic plate

The final two examples in this work are elasto-plastic necking simulations which are well-known [52] to trigger hourglassing for standard elements. First we consider a plate subjected to plane strain conditions (see, e.g., [8, 37, 47, 67, 129, 131]) with length $2L = 53.334$, width $2R = 12.826$ and thickness $t = 1$. Only one fourth (see Figure 5.17) has to be modeled due to symmetry and is meshed with $20 \times 10 \times 1$ elements. Boundary conditions $u(X=0, X, Z) = 0$, $v(X, Y=0, Z) = 0$ and $w(X, Y, Z) = 0$ ensure symmetry and the plane strain state, respectively. Load is applied at the top edge in the form of prescribed displacement $\bar{u} = 7$ applied within 200 load steps[17]. Furthermore, a geometric imperfection in the form of a linear reduction of the width to $\bar{R} = 6.343$ ensures that necking initiates at the lower boundary ($Y = 0$). Figure 5.17 shows the initial geometry, boundary conditions, mesh with refinement in the lower fifth and a typical load displacement curve.

Results of the test are shown in Figure 5.18 which depicts a deformed configuration computed with different finite elements at $\bar{u} = 5.6$. It is clearly visible that H1/E9 shows severe hourglassing. HA1/E12T's spurious mode is less pronounced due to the special nine point integration yet it can still be seen especially along the left edge. If the standard Gauss quadrature was employed the hourglassing patterns would be similar to H1/E9. PG-EAS elements based on F- and E-enhancement do not converge in this test and show severe

[17] This large number of load steps is necessary to capture the hourglass behavior. Elements that do not exhibit spurious modes converge much faster. In particular, the novel H1U/h12 requires 28 load steps.

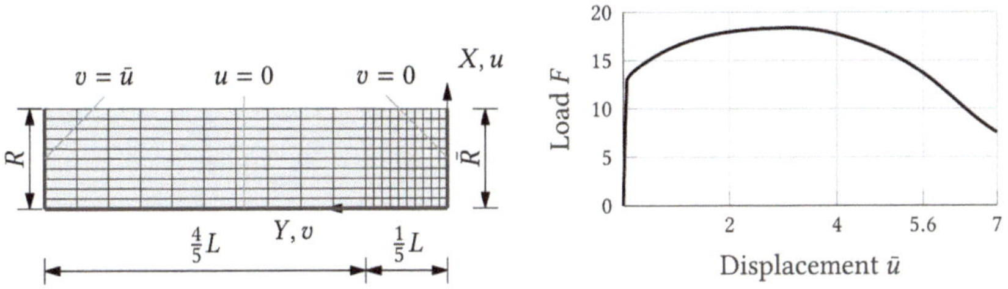

Figure 5.17.: Plane strain elasto-plastic plate. Setup of the example (left) and load displacement curve computed with H1U/h12 (right).

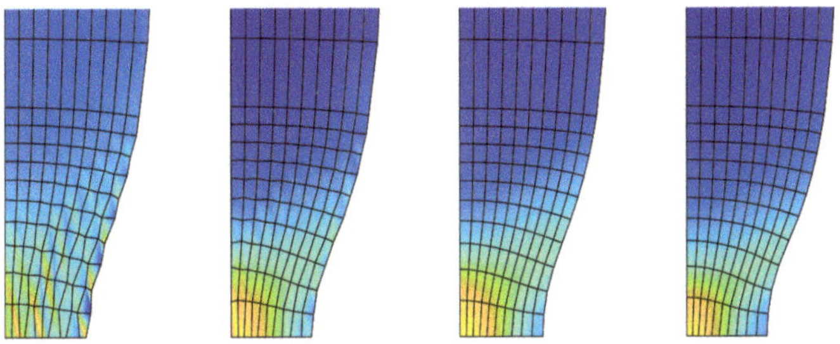

Figure 5.18.: Plane strain elasto-plastic plate. Deformed configuration with distribution of accumulated plastic strain for prescribed displacement $\bar{u} = 5.6$.

hourglassing before the Newton-Raphson scheme fails. Only the mixed pressure EAS element H1/P0E6T and the newly proposed h-enhanced H1U/h12 are hourglassing-free. Especially the latter result is very interesting. To the best knowledge of the authors this is the first EAS element to be hourglassing-free in this test.

5.5.9. Necking of a elasto-plastic circular bar

The final benchmark in this work concerns the necking of an elasto-plastic circular bar (see, e.g., [67, 105, 109, 130–132]). Similar to the previous simulation we consider a cylindric specimen with length $2L = 53.334$ and radius $R = 6.413$ of which only one eight is considered due to symmetry. Figure 5.19 shows the undeformed geometry and the mesh which consists of two parts with 480 elements each. Symmetry boundary conditions $u = 0$, $v = 0$, $w = 0$ apply on the surfaces $X = 0$, $Y = 0$, $Z = 0$, respectively, and load is applied by prescribed displacement $u = \bar{u}$ on the surface $X = L$. Again, the radius is linearly reduced to $\bar{R} = 6.343$ to initiate necking at $X = 0$. The final state $\bar{u} = 7$ is reached within 30 load steps[18] of which half is used up to a displacement of $\bar{u} = 5.6$ and the other half is used to cover the more demanding range up to $\bar{u} = 7.0$.

[18] Only element H1/P0E6T requires a total of 50 load steps due to the hourglassing patterns.

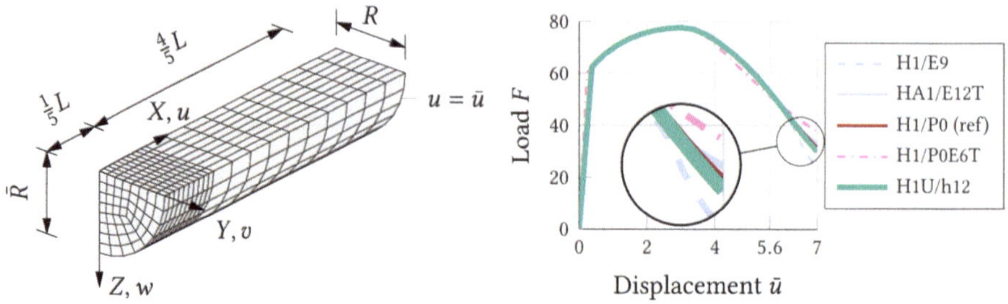

Figure 5.19.: Elasto-plastic circular bar. Setup of the example (left) and load displacement curve for various elements (right).

Figure 5.19 shows the load displacement curves for several finite elements. Element H1/P0 is taken as reference solution since it coincides well with experimental data [133]. Apart from H1/P0E6T the curves are pretty much the same up to $\bar{u} \approx 5.6$ with a maximum load of $F_{\max} = 77.65 \pm 0.1\%$. Afterwards, there are slight differences with H1U/h12 exhibiting the best agreement with H1/P0. Figure 5.20 shows the final configuration for $\bar{u} = 7.0$ computed with various elements. Only H1/P0E6T exhibits spurious hourglassing modes. Thus, to the best knowledge of the authors, H1U/h12 is the only low-order locking-free finite element to be hourglassing-free in both the plane strain (see Section 5.5.8) and circular bar elasto-plastic necking simulation.

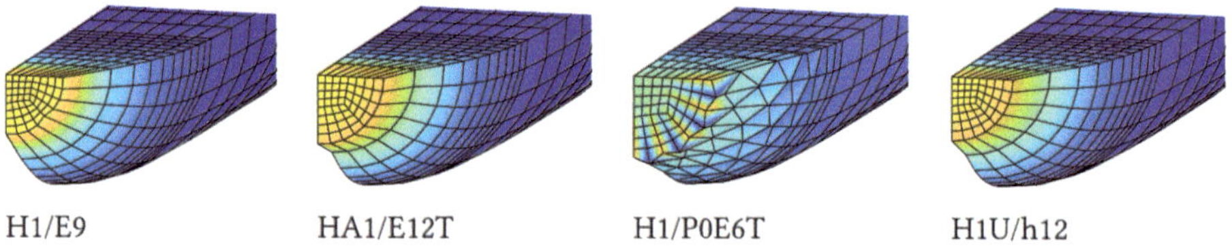

Figure 5.20.: Elasto-plastic circular bar. Deformed configuration with distribution of accumulated plastic strain for prescribed displacement $\bar{u} = 7.0$.

5.6. Conclusion

The present contribution introduced a novel *Petrov-Galerkin EAS element* for large deformation solid mechanics. It is based on the recently published linear elastic framework by the authors [106] and *enhancement of the spatial displacement gradient* (see Schmied [125]). The resulting element has many favorable properties. Besides the standard requirements such as satisfaction of the patch test, objectivity and being locking-free it exhibits two outstanding features. First, the Petrov-Galerkin approach makes the finite element much less sensitive to mesh distortion. This is of great value in practical simulations since it could reduce the effort needed to mesh complex geometries. Second, the finite element is

free from spurious modes in all typically critical examples known to trigger hourglassing for existing finite elements. In fact it is to the best knowledge of the authors the *only* locking-free low-order mixed finite element to be free from hourglassing under compression, tension and for elasto-plastic simulations. Moreover, we postulated a hypothesis on why instabilities arise for all previously proposed EAS elements and other related mixed methods. We supposed that the transmission of well-known checkerboard modes to the enhanced modes is at the core of the problem and showed that the newly proposed approach circumvents the issue.
Two further novelties in the present work concern a slight modification of the higher order enhanced modes and the MIP-method first used for Petrov-Galerkin elements in this contribution. The former reduced the numerical effort to get the ansatz functions and the latter increased robustness of the approach in the Newton-Raphson scheme. Finally, we tested the novel element in a plethora of examples and showed that it outperforms the standard elements used for comparison.

Future work could first focus on tackling the spurious behavior in the mesh distortion test for non-standard meshes mentioned in Remark 5.6. Furthermore, despite the progress made with the MIP method, it would be interesting to further increase the element's robustness to achieve the same performance as assumed stress elements in that regard. Naturally, it would also be valuable to find a proof for the hypothesis on stability of the element. However, we believe that it would be of the utmost interest to apply the methods developed herein to shell elements.

Appendix to Chapter 5

5.A. Higher order incompatible modes

In this appendix we cover analytic solutions for higher order displacement modes in isotropic linear elasticity. These solutions inspire the modified higher order enhanced modes employed in Sections 5.3.2.3 and 5.3.2.4. The benefit of these modes is that they allow to circumvent the tedious orthogonalization required for the element by Pfefferkorn and Betsch [106]. Nadler and Rubin [96] propose a set of higher order displacement modes and use it to design their Cosserat point element. Here we consider similar displacement modes of the form

$$\mathbf{u} = \begin{bmatrix} xyz \\ 0 \\ 0 \end{bmatrix} + \frac{\nu}{6} \begin{bmatrix} 0 \\ -3y^2z + z^3 \\ -3yz^2 + y^3 \end{bmatrix}, \tag{5.74}$$

where $\mathbf{x} = [x, y, z]^{\mathrm{T}}$ are physical coordinates and ν is the Poisson's ratio. The standard relations yield the corresponding strain and stress field given by

$$\varepsilon = \frac{1}{2}\begin{bmatrix} 2yz & xz & xy \\ xz & 0 & 0 \\ xy & 0 & 0 \end{bmatrix} - \nu \begin{bmatrix} 0 & 0 & 0 \\ 0 & yz & 0 \\ 0 & 0 & yz \end{bmatrix}, \qquad \sigma = \mu \begin{bmatrix} 2yz(\nu+1) & xz & xy \\ xz & 0 & 0 \\ xy & 0 & 0 \end{bmatrix}, \tag{5.75}$$

which have a particularly simple form in comparison to the modes proposed by Nadler and Rubin [96]. Since the ansatz for the displacement already contains the trilinear term xyz (see Section 5.3.2.2), only the second part of (5.74) has to be added in the sense of incompatible modes. Thus, the enhanced field of the higher order enhanced modes should have the form of the second matrix in (5.75) which inspired the fields in (5.34) and (5.37). Figure 5.21 shows the respective parts of the displacement field given in (5.74). Unfortunately, (5.74) cannot directly be transferred to the skew coordinate frame in case of distorted meshes. However, it seems that using (5.34) and (5.36) yields good results even in distorted meshes, that is, when the skew frame is not orthogonal.

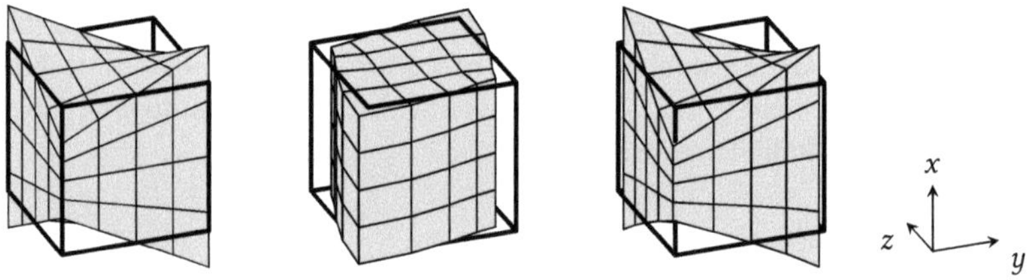

Figure 5.21.: Higher order analytic displacement mode. Trilinear (left), enhanced (middle) and total (right) displacement field (5.74) scaled with 0.5, 2.0 and 0.5, respectively. Computed with $\nu = 0.3$ for a block $\mathcal{B}_0 = [-1, +1]^3$ (thick black line).

6. Conclusion

This final chapter of the present cumulative dissertation summarizes the key findings of the four publications reproduced in Chapter 2 to 5. Furthermore, an outlook to possible further research is given. We refer to the conclusions of the respective chapters for more thorough summaries.

6.1. Summary

The present work gave a comprehensive overview of the state of the art EAS method and introduced three major improvements in order to overcome the issues listed in Section 1.1.

Chapter 2 reproduced [104] and thoroughly covered the history of the EAS method. Furthermore, some minor improvements and novel proofs concerning the patch test have been proposed. The first major improvement was introduced in Chapter 3 (reproducing [109]) in the form of the MIP method for EAS elements. This simple yet effective modification greatly increases the robustness of EAS elements in many examples. Thus, it also improves the numerical efficiency of EAS elements. On top of that, Chapter 3 extends the MIP method to general material models without the requirement of an inverse stress-strain relation, making it easily applicable to a wide range of problems. The second issue studied in this work concerned mesh distortion sensitivity of EAS elements. It was addressed for linear elastic problems in Chapter 4, which reproduced [106]. Therein, we revisited MacNeal's theorem, which essentially states, that any element with a symmetric stiffness matrix must either fail the patch test or be sensitive to mesh distortion. Consequently, a novel Petrov-Galerkin EAS framework (with unsymmetric stiffness matrix) was introduced in Chapter 4 to overcome this limitation. Three design conditions which allow the construction of optimal EAS elements, were established in a next step. The novel framework allows to construct low-order EAS elements, that simultaneously pass the patch test and are mesh distortion insensitive. Finally, the work [108] was reproduced in Chapter 5. The novel EAS framework presented there does not only overcome the third issue of hourglassing instabilities by relying on enhancement of the spatial displacement gradient, but also employs the approaches from Chapters 3 and 4. Thus, the novel element is robust, mesh distortion insensitive and hourglassing-free as confirmed in many numerical investigations.

All in all, the new EAS methods presented in this thesis can be confidently applied in practical simulations and greatly improves upon existing results.

6.2. Outlook

There are still some possibilities to further improve the current EAS framework as mentioned in Section 5.6. It would be worthwhile to address the spurious behavior in the mesh distortion test, if the non-standard meshes proposed in Section 4.5.4 are used, even though a simple fix with mesh-refinement already exists. Another promising line of research concerns robustness. Although it was possible to substantially increase robustness of EAS elements with the MIP method (see Chapter 3), it would still be of great practical value to further improve robustness of EAS elements and meet the performance of assumed stress elements, which are still superior in that regard.

However, an extension of the method to other problems seems even more promising. So far, the Petrov-Galerkin EAS method has (only) been applied in static problems of solid mechanics, for which it proved to work well. A first extension could concern structural finite elements, in particular, shell and solid-shell elements. The present approach might be capable of alleviating transverse shear locking even for distorted meshes. Another limitation of the present approach is its restriction to static problems. Especially an extension to dynamical simulations in combination with structure-preserving integration schemes could be interesting, since the issue of non-symmetric tangential matrices of structure-preserving integrators would not cause additional disadvantages. Such a dynamical extension is currently under investigation by the author and his supervisors. A final extension of the method to other problems concerns multi physics. It can be expected that the methods at hand also improves results in such simulations.

Another restriction of the current work is its limitation to the EAS method. It was chosen since it is probably the most widely applied completely locking-free mixed finite element method. However, the schemes presented in this thesis could also be applied to other types of mixed elements and improve their performance.

Declaration of authorship

The initials Robin Pfefferkorn (RP), Peter Betsch (PB), Simon Bieber (SB), Bastian Oesterle (BO) and Manfred Bischoff (MB) are used below to mark the contributions of the respective authors to the publications.

RESEARCH ARTICLE

WILEY

On transformations and shape functions for enhanced assumed strain elements

Robin Pfefferkorn | Peter Betsch

1 | INTRODUCTION

Citation:

Pfefferkorn R and Betsch P. “On Transformations and Shape Functions for Enhanced Assumed Strain Elements”. In: *Int J Numer Meth Eng.* 120(2): 231–261, 2019. DOI: `10.1002/nme.6133`

Declaration:

PB and RP proposed the novel transformation for the incompatible and compatible part of the deformation gradient, respectively. RP developed the novel proofs for the patch test. RP implemented and verified the novel methods. Moreover, RP wrote the manuscript supported by PB and the final work was reviewed by both authors.

RESEARCH ARTICLE

WILEY

Improving efficiency and robustness of enhanced assumed strain elements for nonlinear problems

Robin Pfefferkorn | Simon Bieber | Bastian Oesterle | Manfred Bischoff | Peter Betsch

1 | INTRODUCTION

Citation:

Pfefferkorn R, Bieber S, Oesterle B, Bischoff M, and Betsch P. “Improving Efficiency and Robustness of EAS Elements for Nonlinear Problems”. In: *Int J Numer Meth Eng.* 122(8): 1911–1939, 2021. DOI: `10.1002/nme.6605`

Declaration:

SB proposed the idea to extend the MIP method to EAS elements. SB and RP worked together on the theory. RP proposed the novel inverse stress-strain relation. Both, SB and RP, implemented the MIP extensions for 2D EAS elements and RP also for 3D elements. SB and RP worked on the numerical investigations together. The manuscript was written by SB and RP and reviewed by all authors (RP, SB, BO, MB, PB).

RESEARCH ARTICLE

WILEY

Mesh distortion insensitive and locking-free Petrov–Galerkin low-order EAS elements for linear elasticity

Robin Pfefferkorn | Peter Betsch

Abstract

One of the most successful mixed finite element methods in solid mechanics is the enhanced assumed strain (EAS) method developed by Simo and Rifai in 1990. However, one major drawback of EAS elements is the highly mesh dependent accuracy. In fact, it can be shown that not only EAS elements, but every finite element with a symmetric stiffness matrix must either fail the patch test or be sensitive to mesh distortion in bending problems (higher order displacement modes) if the shape of the element is arbitrary. This theorem was established by MacNeal in 1992. In the present work we propose a novel Petrov–Galerkin approach for the EAS method, which is equivalent to the standard EAS method in case of regular meshes. However, in case of distorted meshes, it allows to overcome the mesh-distortion sensitivity without loosing other advantages of the EAS method. Three design conditions established in this work facilitate the construction of the element which does not only fulfill the patch test but is also exact in many bending problems regardless of mesh distortion and has an exceptionally high coarse mesh accuracy. Consequently, high quality demands on mesh topology might be relaxed.

1 | INTRODUCTION

In the early days of the finite element method it was soon discovered that *low-order displacement-based* elements severely underestimate displacements under many circumstances such as in bending dominated problems and the incompressible limit. This phenomenon was termed *locking* and prohibits reasonable utilization of low-order displacement-based elements in engineering applications (cf. MacNeal[1]). Thus, a plethora of remedies has been developed since the 1960s which can essentially be grouped into three main categories: higher order methods, reduced integration with stabilization, and mixed finite elements. All of these remedies lead to substantially improved finite elements and some of each category are available in commercial software.

Despite the tremendous effort put into developing new finite elements and enhancing their performance, there have been hardly any major breakthroughs in classical methods since the mid 1990s (with the exemption of the

Citation:
Pfefferkorn R and Betsch P. "Mesh Distortion Insensitive and Locking-Free Petrov-Galerkin Low-Order EAS Elements for Linear Elasticity". In: *Int J Numer Meth Eng.* 122(23): 6924–6954, 2021. DOI: `10.1002/nme.6817`

Declaration:
RP developed the theory for the novel Petrov-Galerkin EAS element and the corresponding design conditions. Moreover, RP implemented the methods and was responsible for the numerical investigations and verifications. RP wrote the manuscript supported by PB. Both authors (RP, PB) conducted the review of the article together.

RESEARCH ARTICLE

WILEY

Hourglassing- and locking-free mesh distortion insensitive Petrov–Galerkin EAS element for large deformation solid mechanics

Robin Pfefferkorn | Peter Betsch

Abstract

We present a novel geometrically nonlinear EAS element with several desirable features. First, a Petrov–Galerkin ansatz significantly improves the element's performance in distorted meshes without loosing the simple strain-driven format. Second, the recently proposed mixed integration point strategy is employed to improve the element's robustness in the Newton–Raphson scheme. Finally and most importantly, we enhance the spatial displacement gradient instead of the usually modified deformation gradient. This allows to construct an element without the well-known spurious instabilities in compression and tension as shown numerically and supported by a corresponding hypothesis. All in all, this leads to a robust, stable, locking-free, and mesh distortion insensitive finite element successfully applied in a wide range of examples.

1 | INTRODUCTION

Countless mixed finite elements have been developed throughout the last 50 years with the goal to cure *locking* and other defects of low-order finite elements. One of the probably most successful element groups are *enhanced assumed strain* (EAS) elements due to their simplicity and strain driven format which allows straightforward implementation of complex material models. The EAS framework has first been proposed in the early 1990s by Simo and Rifai[1] and Simo and Armero[2] for linear and nonlinear problems, respectively, and is a mathematically sound successor of the earlier proposed incompatible mode models. Since then, the EAS method has been implemented into several commercial codes and has not only been successfully applied in solid mechanics but also to model shell structures, diffusion problems and even fracture problems. Despite the tremendous effort put in development of EAS elements there are still some open issues. Pfefferkorn and Betsch discuss three of which: robustness in the Newton–Raphson scheme, mesh-distortion sensitivity and hourglassing instabilities. All three are addressed in this contribution.

Citation:
Pfefferkorn R and Betsch P. "Hourglassing- and Locking-Free Mesh Distortion Insensitive Petrov-Galerkin EAS Element for Large Deformation Solid Mechanics". In: *Int J Numer Meth Eng.* 124(6): 1307–1343, 2023. DOI: `10.1002/nme.7166`

Declaration:
RP proposed the novel enhancement approach and further theory for the newly proposed EAS element. Implementation and extensive numerical investigations were also due to RP. Moreover, RP wrote the manuscript. Finally, the article was reviewed by RP and PB.

Publications and talks

Peer-reviewed publications

- **Pfefferkorn R** and Betsch P. "On Transformations and Shape Functions for Enhanced Assumed Strain Elements". In: *Int J Numer Meth Eng.* 120(2): 231–261, 2019. DOI: 10.1002/nme.6133.
- **Pfefferkorn R** and Betsch P. "Extension of the Enhanced Assumed Strain Method Based on the Structure of Polyconvex Strain-Energy Functions". In: *Int J Numer Meth Eng.* 121(8): 1695–1737, 2020. DOI: 10.1002/nme.6284.
- **Pfefferkorn R** and Betsch P. "Mesh Distortion Insensitive and Locking-Free Petrov-Galerkin Low-Order EAS Elements for Linear Elasticity". In: *Int J Numer Meth Eng.* 122(23): 6924–6954, 2021. DOI: 10.1002/nme.6817.
- **Pfefferkorn R**, Bieber S, Oesterle B, Bischoff M, and Betsch P. "Improving Efficiency and Robustness of EAS Elements for Nonlinear Problems". In: *Int J Numer Meth Eng.* 122(8): 1911–1939, 2021. DOI: 10.1002/nme.6605.
- **Pfefferkorn R** and Betsch P. "Hourglassing- and Locking-Free Mesh Distortion Insensitive Petrov-Galerkin EAS Element for Large Deformation Solid Mechanics". In: *Int J Numer Meth Eng.* 124(6): 1307–1343, 2023. DOI: 10.1002/nme.7166.

Other publications

- **Pfefferkorn R** and Betsch P. "Transformations and Enhancement Approaches for the Large Deformation EAS Method". In: *Proc Appl Math Mech.* 19(1): e201900135, 2019. DOI: 10.1002/pamm.201900135.
- **Pfefferkorn R** and Betsch P. "Open Issues on the EAS Method and Mesh Distortion Insensitive Locking-Free Low-Order Unsymmetric EAS Elements". In: *14th WCCM-ECCOMAS Congress 2020.* Paris, France, 2021. DOI: 10.23967/wccm-eccomas.2020.264.
- Hille M, **Pfefferkorn R**, and Betsch P. "Locking-Free Mixed Finite Element Methods and their Spurious Hourglassing Patterns". In: *Current Trends and Open Problems in Computational Mechanics.* Ed. by Aldakheel F, Hudobivnik B, Soleimani M, Wessels H, Weißenfels C, and Marino M. Cham: Springer, 2022, pp. 181–194. DOI: 10.1007/978-3-030-87312-7_19.

Conference talks

- **Pfefferkorn R** and Betsch P. *Extension of the Nonlinear EAS Method Based on the Structure of Polyconvex Strain-Energy Functions.* ECCOMAS Thematic Conference - Modern Finite Element Technologies (MFET). Bad Honnef, Germany, July 1–3, 2019.
- **Pfefferkorn R** and Betsch P. *Transformations and Enhancement Approaches for the Large Deformation EAS Method.* 90th GAMM Annual Meeting. Vienna, Austria, Feb. 18–22, 2019.
- **Pfefferkorn R** and Betsch P. *High Performance Mesh Distortion Insensitive Unsymmetric EAS Elements for Linear Elasticity.* 90th GAMM Annual Meeting. Kassel, Germany (virtual congress), Mar. 15–19, 2021.
- **Pfefferkorn R** and Betsch P. *Open Issues on and Extensions to the Large Deformation EAS Method.* 7th European Community on Computational Methods in Applied Sciences and Engineering (ECCOMAS) & 14th World Congress on Computational Mechanics (WCCM). Paris, France (virtual congress), Jan. 11–15, 2021.
- **Pfefferkorn R** and Betsch P. *Petrov-Galerkin EAS Elements for Mesh-Distortion Insensitive FE-Simulations in Solid Mechanics.* 6th ECCOMAS Young Investigators Conference (YIC). Valencia, Spain (virtual congress), July 7–9, 2021.
- **Pfefferkorn R**, Bieber S, Oesterle B, Bischoff M, and Betsch P. *Increasing Robustness of Large Deformation EAS Elements in Newton-Raphson Iterations.* 7th European Community on Computational Methods in Applied Sciences and Engineering (ECCOMAS) & 14th World Congress on Computational Mechanics (WCCM). Paris, France (virtual congress), Jan. 11–15, 2021.
- **Pfefferkorn R** and Betsch P. *Hourglassing-Free Petrov Galerkin Enhanced Assumed Strain Finite Elements Insensitive to Mesh-Distortion for Nonlinear Solid Mechanics.* 15th World Congress on Computational Mechanics (WCCM) & 8th Asian Pacific Congress on Computational Mechanics (APCOM). Yokohama, Japan (virtual congress), July 31–Aug. 5, 2022.
- **Pfefferkorn R** and Betsch P. *Mesh Distortion Insensitive Hourglassing-Free Petrov-Galerkin EAS Elements for Large Deformation Solid Mechanics.* 92th GAMM Annual Meeting. Aachen, Germany, Aug. 15–19, 2022.
- **Pfefferkorn R**, Gil AJ, Ortigosa R, and Betsch P. *Energy-Momentum Preserving Time Integration Schemes for Petrov-Galerkin EAS Mixed Finite Elements.* 8th European Congress on Computational Methods in Applied Sciences and Engineering (ECCOMAS). Oslo, Norway, June 5–9, 2022.

List of Figures

List of Tables

Bibliography

[1] Alves de Sousa RJ, Natal Jorge RM, Fontes Valente RA, and César de Sá JMA. "A New Volumetric and Shear Locking-free 3D Enhanced Strain Element". In: *Eng Comput.* 20(7): 896–925, 2003. DOI: 10.1108/02644400310502036.

[2] Andelfinger U. "Untersuchungen zur Zuverlässigkeit hybrid-gemischter Finiter Elemente für Flächentragwerke". PhD thesis. Stuttgart: University of Stuttgart, 1991.

[3] Andelfinger U and Ramm E. "EAS-elements for Two-Dimensional, Three-Dimensional, Plate and Shell Structures and Their Equivalence to HR-elements". In: *Int J Numer Meth Eng.* 36(8): 1311–1337, 1993. DOI: 10.1002/nme.1620360805.

[4] Andelfinger U, Ramm E, and Roehl D. "2D and 3D Enhanced Assumed Strain Elements and Their Application in Plasticity". In: *Proc. of the Third International Conference on Computational Plasticity (COMPLAS).* Ed. by Owen DRJ, Oñate E, and Hinton E. Barcelona, Spain: Pineridge Press, Apr. 1992.

[5] Angoshtari A, Faghih Shojaei M, and Yavari A. "Compatible-Strain Mixed Finite Element Methods for 2D Compressible Nonlinear Elasticity". In: *Comput Methods Appl Mech Eng.* 313: 596–631, 2017. DOI: 10.1016/j.cma.2016.09.047.

[6] Areias PMA, César de Sá JMA, António CAC, and Fernandes AA. "Analysis of 3D Problems Using a New Enhanced Strain Hexahedral Element". In: *Int J Numer Meth Eng.* 58(11): 1637–1682, 2003. DOI: 10.1002/nme.835.

[7] Armero F. *A Modal Analysis of Finite Deformation Enhanced Strain Finite Elements.* UCB/SEMM Report 96/03. Berkeley, CA: University of California, 1996.

[8] Armero F. "On the Locking and Stability of Finite Elements in Finite Deformation Plane Strain Problems". In: *Comput Struct.* 75(3): 261–290, 2000. DOI: 10.1016/s0045-7949(99)00136-4.

[9] Armero F and Linder C. "New Finite Elements with Embedded Strong Discontinuities in the Finite Deformation Range". In: *Comput Methods Appl Mech Eng.* 197(33-40): 3138–3170, 2008. DOI: 10.1016/j.cma.2008.02.021.

[10] Arunakirinathar K and Reddy BD. "Further Results for Enhanced Strain Methods with Isoparametric Elements". In: *Comput Methods Appl Mech Eng.* 127(1): 127–143, 1995. DOI: 10.1016/0045-7825(95)00845-0.

[11] Auricchio F, Beirão da Veiga L, Lovadina C, and Reali A. "Stability of Some Finite Element Methods for Finite Elasticity Problems". In: *Mixed Finite Element Technologies.* Ed. by Carstensen C and Wriggers P. Vol. 509. CISM International Centre for Mechanical Sciences. Vienna: Springer, 2009, pp. 179–206. DOI: 10.1007/978-3-211-99094-0_6.

[12] Auricchio F, Beirão da Veiga L, Lovadina C, and Reali A. "A Stability Study of Some Mixed Finite Elements for Large Deformation Elasticity Problems". In: *Comput Methods Appl Mech Eng.* 194(9-11): 1075–1092, 2005. DOI: 10.1016/j.cma.2004.06.014.

[13] Başar Y and Ding Y. "Finite-Rotation Shell Elements for the Analysis of Finite-Rotation Shell Problems". In: *Int J Numer Meth Eng.* 34(1): 165–169, 1992. DOI: 10.1002/nme.1620340109.

[14] Belytschko T and Bindeman LP. "Assumed Strain Stabilization of the Eight Node Hexahedral Element". In: *Comput Methods Appl Mech Eng.* 105(2): 225–260, 1993. DOI: 10.1016/0045-7825(93)90124-G.

[15] Belytschko T, Liu WK, Moran B, and Elkhodary KI. *Nonlinear Finite Elements for Continua and Structures.* 2nd ed. Chichester: Wiley, 2014.

[16] Belytschko T, Ong JSJ, Liu WK, and Kennedy JM. "Hourglass Control in Linear and Nonlinear Problems". In: *Comput Methods Appl Mech Eng.* 43(3): 251–276, 1984. DOI: 10.1016/0045-7825(84)90067-7.

[17] Betsch P, Gruttmann F, and Stein E. "A 4-Node Finite Shell Element for the Implementation of General Hyperelastic 3D-elasticity at Finite Strains". In: *Comput Methods Appl Mech Eng.* 130(1): 57–79, 1996. DOI: 10.1016/0045-7825(95)00920-5.

[18] Betsch P, Franke M, and Janz A. "Polyconvexity and the Design of Mixed Finite Elements". In: *Proc. ECCOMAS Thematic Conference on Modern Finite Element Technologies (MFET).* Bad Honnef, Germany, Aug. 2017.

[19] Betsch P, Janz A, and Hesch C. "A Mixed Variational Framework for the Design of Energy-Momentum Schemes Inspired by the Structure of Polyconvex Stored Energy Functions". In: *Comput Methods Appl Mech Eng.* 335: 660–696, 2018. DOI: 10.1016/j.cma.2018.01.013.

[20] Bieber S, Auricchio F, Reali A, and Bischoff M. "Artificial Instabilities of Finite Elements for Finite Elasticity: Analysis and Remedies". In: *Int J Numer Meth Eng.*, 2022. Submitted. Preprint available at: DOI: 10.31224/2604.

[21] Bischoff M, Ramm E, and Braess D. "A Class of Equivalent Enhanced Assumed Strain and Hybrid Stress Finite Elements". In: *Comput Mech.* 22(6): 443–449, 1999. DOI: 10.1007/s004660050378.

[22] Bischoff M and Romero I. "A Generalization of the Method of Incompatible Modes". In: *Int J Numer Meth Eng.* 69(9): 1851–1868, 2007. DOI: 10.1002/nme.1830.

[23] Boffi D, Brezzi F, and Fortin M. *Mixed Finite Element Methods and Applications.* Vol. 44. Springer Series in Computational Mathematics. Berlin: Springer, 2013. DOI: 10.1007/978-3-642-36519-5.

[24] Bonet J, Gil AJ, and Ortigosa R. "A Computational Framework for Polyconvex Large Strain Elasticity". In: *Comput Methods Appl Mech Eng.* 283: 1061–1094, 2015. DOI: 10.1016/j.cma.2014.10.002.

[25] Bonet J and Wood RD. *Nonlinear Continuum Mechanics for Finite Element Analysis.* 2nd ed. Cambridge: Cambridge University Press, 2008. DOI: 10.1017/cbo9780511755446.

[26] Braess D. *Finite Elements: Theory, Fast Solvers, and Applications in Elasticity Theory.* 3rd ed. Cambridge: Cambridge University Press, 2007. DOI: 10.1017/CBO9780511618635.

[27] Brink U and Stein E. "On Some Mixed Finite Element Methods for Incompressible and Nearly Incompressible Finite Elasticity". In: *Comput Mech.* 19(1): 105–119, 1996. DOI: 10.1007/BF02824849.

[28] Büchter N, Ramm E, and Roehl D. "Three-Dimensional Extension of Non-Linear Shell Formulation Based on the Enhanced Assumed Strain Concept". In: *Int J Numer Meth Eng.* 37(15): 2551–2568, 1994. DOI: 10.1002/nme.1620371504.

[29] Caseiro JF, Alves de Sousa RJ, and Valente RAF. "A Systematic Development of EAS Three-Dimensional Finite Elements for the Alleviation of Locking Phenomena". In: *Finite Elem Anal Des.* 73: 30–41, 2013. DOI: 10.1016/j.finel.2013.05.006.

[30] Caylak I and Mahnken R. "Mixed Finite Element Formulations with Volume Bubble Functions for Triangular Elements". In: *Comput Struct.* 89(21): 1844–1851, 2011. DOI: 10.1016/j.compstruc.2011.07.004.

[31] Caylak I and Mahnken R. "Stabilization of Mixed Tetrahedral Elements at Large Deformations". In: *Int J Numer Meth Eng.* 90(2): 218–242, 2012. DOI: 10.1002/nme.3320.

[32] Cen S, Zhou PL, Li CF, and Wu CJ. "An Unsymmetric 4-Node, 8-DOF Plane Membrane Element Perfectly Breaking through MacNeal's Theorem". In: *Int J Numer Meth Eng.* 103(7): 469–500, 2015. DOI: 10.1002/nme.4899.

[33] Clough RW. "Early History of the Finite Element Method from the View Point of a Pioneer". In: *Int J Numer Meth Eng.* 60(1): 283–287, 2004. DOI: 10.1002/nme.962.

[34] Cottrell JA, Hughes TJR, and Bazilevs Y. *Isogeometric Analysis: Toward Integration of CAD and FEA.* Chichester: Wiley, 2009. DOI: 10.1002/9780470749081.

[35] Crisfield MA and Moita GF. "A Co-Rotational Formulation for 2-D Continua Including Incompatible Modes". In: *Int J Numer Meth Eng.* 39(15): 2619–2633, 1996. DOI: 10.1002/(SICI)1097-0207(19960815)39:15<2619::AID-NME969>3.0.CO;2-N.

[36] Crisfield MA, Moita GF, Jelenić G, and Lyons LPR. "Enhanced Lower-Order Element Formulations for Large Strains". In: *Comput Mech.* 17(1): 62–73, 1995. DOI: 10.1007/BF00356479.

[37] De Souza Neto EA, Perić D, Huang GC, and Owen DRJ. "Remarks on the Stability of Enhanced Strain Elements in Finite Elasticity and Elastoplasticity". In: *Commun Numer Meth Eng.* 11(11): 951–961, 1995. DOI: 10.1002/cnm.1640111109.

[38] Di S and Ramm E. "On Alternative Hybrid Stress 2d and 3d Elements". In: *Eng Comput.* 11(1): 49–68, 1994. DOI: 10.1108/02644409410799155.

[39] Djoko JK, Lamichhane BP, Reddy BD, and Wohlmuth BI. "Conditions for Equivalence between the Hu-Washizu and Related Formulations, and Computational Behavior in the Incompressible Limit". In: *Comput Methods Appl Mech Eng.* 195(33): 4161–4178, 2006. DOI: 10.1016/j.cma.2005.07.018.

[40] Flanagan DP and Belytschko T. "A Uniform Strain Hexahedron and Quadrilateral with Orthogonal Hourglass Control". In: *Int J Numer Meth Eng.* 17(5): 679–706, 1981. DOI: 10.1002/nme.1620170504.

[41] Fontes Valente RA, Alves de Sousa RJ, and Natal Jorge RM. "An Enhanced Strain 3D Element for Large Deformation Elastoplastic Thin-Shell Applications". In: *Comput Mech.* 34(1): 38–52, 2004. DOI: 10.1007/s00466-004-0551-7.

[42] Fraeijs de Veubeke B. "Displacement and Equilibrium Models in the Finite Element Method". In: *Stress Analysis.* Ed. by Zienkiewicz OC and Holister GS. Chichester: John Wiley & Sons, 1965, pp. 145–197.

[43] Garcea G. "Mixed Formulation in Koiter Analysis of Thin-Walled Beams". In: *Comput Methods Appl Mech Eng.* 190(26): 3369–3399, 2001. DOI: 10.1016/S0045-7825(00)00268-1.

[44] Garcea G, Salerno G, and Casciaro R. "Extrapolation Locking and Its Sanitization in Koiter's Asymptotic Analysis". In: *Comput Methods Appl Mech Eng.* 180(1): 137–167, 1999. DOI: 10.1016/S0045-7825(99)00053-5.

[45] Garcea G, Trunfio GA, and Casciaro R. "Mixed Formulation and Locking in Path-Following Nonlinear Analysis". In: *Comput Methods Appl Mech Eng.* 165(1): 247–272, 1998. DOI: 10.1016/S0045-7825(98)00068-1.

[46] Glaser S and Armero F. *Recent Developments in the Formulation of Assumed Enhanced Strain Finite Elements for Finite Deformation Problems.* UCB/SEMM Report 95/13. Berkeley, CA: University of California, 1995.

[47] Glaser S and Armero F. "On the Formulation of Enhanced Strain Finite Elements in Finite Deformations". In: *Eng Comput.* 14(7): 759–791, 1997. DOI: 10.1108/02644409710188664.

[48] Gonzalez O and Stuart AM. *A First Course in Continuum Mechanics.* Cambridge: Cambridge University Press, 2008. DOI: 10.1017/CBO9780511619571.

[49] Hartmann S and Neff P. "Polyconvexity of Generalized Polynomial-Type Hyperelastic Strain Energy Functions for near-Incompressibility". In: *Int J Solids Struct.* 40(11): 2767–2791, 2003. DOI: 10.1016/S0020-7683(03)00086-6.

[50] Haupt P. *Continuum Mechanics and Theory of Materials.* 2nd ed. Berlin: Springer, 2002. DOI: 10.1007/978-3-662-04109-3.

[51] Herrmann LR. “Elasticity Equations for Incompressible and Nearly Incompressible Materials by a Variational Theorem”. In: *AIAA J.* 3(10): 1896–1900, 1965. DOI: 10.2514/3.3277.

[52] Hille M, Pfefferkorn R, and Betsch P. “Locking-Free Mixed Finite Element Methods and their Spurious Hourglassing Patterns”. In: *Current Trends and Open Problems in Computational Mechanics.* Ed. by Aldakheel F, Hudobivnik B, Soleimani M, Wessels H, Weißenfels C, and Marino M. Cham: Springer, 2022, pp. 181–194. DOI: 10.1007/978-3-030-87312-7_19.

[53] Holzapfel GA. *Nonlinear Solid Mechanics: A Continuum Approach for Engineering.* Chichester: Wiley, 2000.

[54] Hu S, Xu J, Liu X, and Yan M. “Eight-Node Nonconforming Hexahedral Element Based on Reverse Adjustment to Patch Test for Solids, Beams, Plates, and Shells”. In: *Int J Numer Meth Eng.* 119(5): 361–382, 2019. DOI: 10.1002/nme.6053.

[55] Hu S, Xu J, Liu X, and Yan M. “Reverse Adjustment to Patch Test and Two 8-Node Hexahedral Elements”. In: *Eur J Mech A/Solids.* 73: 430–436, 2019. DOI: 10.1016/j.euromechsol.2018.10.010.

[56] Huang J, Cen S, Li Z, and Li CF. “An Unsymmetric 8-Node Hexahedral Solid-Shell Element with High Distortion Tolerance: Linear Formulations”. In: *Int J Numer Meth Eng.* 116(12-13): 759–783, 2018. DOI: 10.1002/nme.5945.

[57] Huang YQ and Li QS. “Four-Node Incompatible Plane and Axisymmetric Elements with Quadratic Completeness in the Physical Space”. In: *Int J Numer Meth Eng.* 61(10): 1603–1624, 2004. DOI: 10.1002/nme.1122.

[58] Huang YQ, Yang YF, Wang JZ, Liu XC, and Chen HB. “Unsymmetric Extensions of Wilson’s Incompatible Four-Node Quadrilateral and Eight-Node Hexahedral Elements”. In: *Int J Numer Meth Eng.* 123(1): 101–127, 2021. DOI: 10.1002/nme.6849.

[59] Huang Y, Huan Y, and Chen H. “An Incompatible and Unsymmetric Four-Node Quadrilateral Plane Element with High Numerical Performance”. In: *Int J Numer Meth Eng.* 121(15): 3382–3396, 2020. DOI: 10.1002/nme.6363.

[60] Hughes TJR. *The Finite Element Method: Linear Static and Dynamic Finite Element Analysis.* Mineola, NY: Dover Publ., 2000.

[61] Irons BM. “Engineering Applications of Numerical Integration in Stiffness Methods”. In: *AIAA J.* 4(11): 2035–2037, 1966. DOI: 10.2514/3.3836.

[62] Jabareen M and Rubin MB. “A 3-D Brick Cosserat Point Element (CPE) for Nonlinear Elasticity”. In: *Computational and Experimental Mechanics of Advanced Materials.* Ed. by Silberschmidt VV. Vol. 514. CISM International Centre for Mechanical Sciences. Vienna: Springer, 2010, pp. 83–140. DOI: 10.1007/978-3-211-99685-0_3.

[63] Kasper EP and Taylor RL. “A Mixed-Enhanced Strain Method. Part I: Geometrically Linear Problems”. In: *Comput Struct.* 75(3): 237–250, 2000. DOI: 10.1016/S0045-7949(99)00134-0.

[64] Kasper EP and Taylor RL. “A Mixed-Enhanced Strain Method. Part II: Geometrically Nonlinear Problems”. In: *Comput Struct.* 75(3): 251–260, 2000. DOI: 10.1016/S0045-7949(99)00135-2.

[65] Klinkel S, Gruttmann F, and Wagner W. “A Robust Non-Linear Solid Shell Element Based on a Mixed Variational Formulation”. In: *Comput Methods Appl Mech Eng.* 195(1): 179–201, 2006. DOI: 10.1016/j.cma.2005.01.013.

[66] Klinkel S and Wagner W. “A Geometrical Non-Linear Brick Element Based on the EAS-method”. In: *Int J Numer Meth Eng.* 40(24): 4529–4545, 1997. DOI: 10.1002/(SICI)1097-0207(19971230)40:24<4529::AID-NME271>3.0.CO;2-I.

[67] Korelc J, Šolinc U, and Wriggers P. “An Improved EAS Brick Element for Finite Deformation”. In: *Comput Mech.* 46(4): 641–659, 2010. DOI: 10.1007/s00466-010-0506-0.

[68] Korelc J and Wriggers P. “An Efficient 3D Enhanced Strain Element with Taylor Expansion of the Shape Functions”. In: *Comput Mech.* 19(2): 30–40, 1996. DOI: 10.1007/BF02757781.

[69] Korelc J and Wriggers P. “Consistent Gradient Formulation for a Stable Enhanced Strain Method for Large Deformations”. In: *Eng Comput.* 13(1): 103–123, 1996. DOI: 10.1108/02644409610111001.

[70] Korelc J and Wriggers P. “Improved Enhanced Strain Four-Node Element with Taylor Expansion of the Shape Functions”. In: *Int J Numer Meth Eng.* 40(3): 407–421, 1997. DOI: 10.1002/(SICI)1097-0207(19970215)40:3<407::AID-NME70>3.0.CO;2-P.

[71] Krischok A and Linder C. “On the Enhancement of Low-Order Mixed Finite Element Methods for the Large Deformation Analysis of Diffusion in Solids”. In: *Int J Numer Meth Eng.* 106(4): 278–297, 2016. DOI: 10.1002/nme.5120.

[72] Kuo-Mo H. “Nonlinear Analysis of General Shell Structures by Flat Triangular Shell Element”. In: *Comput Struct.* 25(5): 665–675, 1987. DOI: 10.1016/0045-7949(87)90159-3.

[73] Kuo-Mo H and Yeh-Ren C. “Nonlinear Analysis of Shell Structures by Degenerated Isoparametric Shell Element”. In: *Comput Struct.* 31(3): 427–438, 1989. DOI: 10.1016/0045-7949(89)90390-8.

[74] Lamichhane BP, Reddy BD, and Wohlmuth BI. “Convergence in the Incompressible Limit of Finite Element Approximations Based on the Hu-Washizu Formulation”. In: *Numer Math.* 104(2): 151–175, 2006. DOI: 10.1007/s00211-006-0014-5.

[75] Leonetti L, Liguori F, Magisano D, and Garcea G. “An Efficient Isogeometric Solid-Shell Formulation for Geometrically Nonlinear Analysis of Elastic Shells”. In: *Comput Methods Appl Mech Eng.* 331: 159–183, Apr. 2018. DOI: 10.1016/j.cma.2017.11.025.

[76] Li Z, Cen S, Huang J, and Li CF. "Hyperelastic Finite Deformation Analysis with the Unsymmetric Finite Element Method Containing Homogeneous Solutions of Linear Elasticity". In: *Int J Numer Meth Eng.* 121: 3702–3721, 16 2020. DOI: 10.1002/nme.6378.

[77] Li Z, Cen S, and Li CF. "Extension of the Unsymmetric 8-Node Hexahedral Solid Element US-ATFH8 to Geometrically Nonlinear Analysis". In: *Eng Comput.* 38(8): 3219–3253, 2021. DOI: 10.1108/EC-04-2020-0203.

[78] Li Z, Cen S, Wu CJ, Shang Y, and Li CF. "High-Performance Geometric Nonlinear Analysis with the Unsymmetric 4-Node, 8-DOF Plane Element US-ATFQ4". In: *Int J Numer Meth Eng.* 114(9): 931–954, 2018. DOI: 10.1002/nme.5771.

[79] Liesen J and Mehrmann V. *Lineare Algebra.* Wiesbaden: Springer Spektrum, 2015. DOI: 10.1007/978-3-658-06610-9.

[80] Linder C and Armero F. "Finite Elements with Embedded Strong Discontinuities for the Modeling of Failure in Solids". In: *Int J Numer Meth Eng.* 72(12): 1391–1433, 2007. DOI: 10.1002/nme.2042.

[81] Loikkanen MJ and Irons BM. "An 8-Node Brick Finite Element". In: *Int J Numer Meth Eng.* 20(3): 523–528, 1984. DOI: 10.1002/nme.1620200310.

[82] Lovadina C and Auricchio F. "On the Enhanced Strain Technique for Elasticity Problems". In: *Comput Struct.* 81(8): 777–787, 2003. DOI: 10.1016/S0045-7949(02)00412-1.

[83] MacNeal RH. "A Theorem Regarding the Locking of Tapered Four-Noded Membrane Elements". In: *Int J Numer Meth Eng.* 24(9): 1793–1799, 1987. DOI: 10.1002/nme.1620240913.

[84] MacNeal RH. "On the Limits of Finite Element Perfectability". In: *Int J Numer Meth Eng.* 35(8): 1589–1601, 1992. DOI: 10.1002/nme.1620350804.

[85] MacNeal RH. *Finite Elements: Their Design and Performance.* New York: Dekker, 1994.

[86] MacNeal RH and Harder RL. "A Proposed Standard Set of Problems to Test Finite Element Accuracy". In: *Finite Elem Anal Des.* 1(1): 3–20, 1985. DOI: 10.1016/0168-874X(85)90003-4.

[87] Magisano D, Leonetti L, and Garcea G. "Advantages of the Mixed Format in Geometrically Nonlinear Analysis of Beams and Shells Using Solid Finite Elements". In: *Int J Numer Meth Eng.* 109(9): 1237–1262, 2016. DOI: 10.1002/nme.5322.

[88] Magisano D, Leonetti L, and Garcea G. "How to Improve Efficiency and Robustness of the Newton Method in Geometrically Non-Linear Structural Problem Discretized via Displacement-Based Finite Elements". In: *Comput Methods Appl Mech Eng.* 313: 986–1005, 2017. DOI: 10.1016/j.cma.2016.10.023.

[89] Magisano D, Leonetti L, Madeo A, and Garcea G. "A Large Rotation Finite Element Analysis of 3D Beams by Incremental Rotation Vector and Exact Strain Measure with All the Desirable Features". In: *Comput Methods Appl Mech Eng.* 361: 112811, 2020. DOI: 10.1016/j.cma.2019.112811.

[90] MATLAB. *9.8.0.1323502 (R2020a).* Natick, MA: The MathWorks Inc., 2019.

[91] Mei Y, Hurtado DE, Pant S, and Aggarwal A. "On Improving the Numerical Convergence of Highly Nonlinear Elasticity Problems". In: *Comput Methods Appl Mech Eng.* 337: 110–127, 2018. DOI: 10.1016/j.cma.2018.03.033.

[92] Miehe C. "Aspects of the Formulation and Finite Element Implementation of Large Strain Isotropic Elasticity". In: *Int J Numer Meth Eng.* 37(12): 1981–2004, 1994. DOI: 10.1002/nme.1620371202.

[93] Moita GF. "Non-Linear Finite Element Analysis of Continua with Emphasis on Hyperelasticity." PhD thesis. London: Imperial College, 1994. URL: http://hdl.handle.net/10044/1/7943 (visited on 2022-11-08).

[94] Müller A and Bischoff M. "A Consistent Finite Element Formulation of the Geometrically Non-linear Reissner-Mindlin Shell Model". In: *Arch Comput Meth Eng.* 29(5): 3387–3434, 2022. DOI: 10.1007/s11831-021-09702-7.

[95] Müller-Hoeppe DS, Löhnert S, and Wriggers P. "A Finite Deformation Brick Element with Inhomogeneous Mode Enhancement". In: *Int J Numer Meth Eng.* 78(10): 1164–1187, 2009. DOI: 10.1002/nme.2523.

[96] Nadler B and Rubin MB. "A New 3-D Finite Element for Nonlinear Elasticity Using the Theory of a Cosserat Point". In: *Int J Solids Struct.* 40(17): 4585–4614, 2003. DOI: 10.1016/S0020-7683(03)00210-5.

[97] Nagtegaal JC and Fox DD. "Using Assumed Enhanced Strain Elements for Large Compressive Deformation". In: *Int J Solids Struct.* 33(20): 3151–3159, 1996. DOI: 10.1016/0020-7683(95)00250-2.

[98] Ogden RW. *Non-Linear Elastic Deformations.* Mineola, NY: Dover Publ., 1997.

[99] Ooi ET, Rajendran S, and Yeo JH. "A 20-Node Hexahedron Element with Enhanced Distortion Tolerance". In: *Int J Numer Meth Eng.* 60(15): 2501–2530, 2004. DOI: 10.1002/nme.1056.

[100] Ooi ET, Rajendran S, and Yeo JH. "Remedies to Rotational Frame Dependence and Interpolation Failure of US-QUAD8 Element". In: *Commun Numer Meth Eng.* 24(11): 1203–1217, 2008. DOI: 10.1002/cnm.1026.

[101] Ooi ET, Rajendran S, and Yeo JH. "Extension of Unsymmetric Finite Elements US-QUAD8 and US-HEXA20 for Geometric Nonlinear Analyses". In: *Eng Comput.* 24(4): 407–431, 2007. DOI: 10.1108/02644400710748715.

[102] Pantuso D and Bathe KJ. "On the Stability of Mixed Finite Elements in Large Strain Analysis of Incompressible Solids". In: *Finite Elem Anal Des.* 28(2): 83–104, 1997. DOI: 10.1016/S0168-874X(97)81953-1.

[103] Petersen KB and Pedersen MS. *The Matrix Cookbook.* Lyngby: Technical University of Denmark, 2012. URL: http://www2.imm.dtu.dk/pubdb/p.php?3274 (visited on 2022-11-08).

[104] Pfefferkorn R and Betsch P. "On Transformations and Shape Functions for Enhanced Assumed Strain Elements". In: *Int J Numer Meth Eng.* 120(2): 231–261, 2019. DOI: 10.1002/nme.6133.

[105] Pfefferkorn R and Betsch P. "Extension of the Enhanced Assumed Strain Method Based on the Structure of Polyconvex Strain-Energy Functions". In: *Int J Numer Meth Eng.* 121(8): 1695–1737, 2020. DOI: 10.1002/nme.6284.

[106] Pfefferkorn R and Betsch P. "Mesh Distortion Insensitive and Locking-Free Petrov-Galerkin Low-Order EAS Elements for Linear Elasticity". In: *Int J Numer Meth Eng.* 122(23): 6924–6954, 2021. DOI: 10.1002/nme.6817.

[107] Pfefferkorn R and Betsch P. "Open Issues on the EAS Method and Mesh Distortion Insensitive Locking-Free Low-Order Unsymmetric EAS Elements". In: *14th WCCM-ECCOMAS Congress 2020.* Paris, France, 2021. DOI: 10.23967/wccm-eccomas.2020.264.

[108] Pfefferkorn R and Betsch P. "Hourglassing- and Locking-Free Mesh Distortion Insensitive Petrov-Galerkin EAS Element for Large Deformation Solid Mechanics". In: *Int J Numer Meth Eng.* 124(6): 1307–1343, 2023. DOI: 10.1002/nme.7166.

[109] Pfefferkorn R, Bieber S, Oesterle B, Bischoff M, and Betsch P. "Improving Efficiency and Robustness of EAS Elements for Nonlinear Problems". In: *Int J Numer Meth Eng.* 122(8): 1911–1939, 2021. DOI: 10.1002/nme.6605.

[110] Pian THH and Sumihara K. "Rational Approach for Assumed Stress Finite Elements". In: *Int J Numer Meth Eng.* 20(9): 1685–1695, 1984. DOI: 10.1002/nme.1620200911.

[111] Pian THH and Chen D. "On the Suppression of Zero Energy Deformation Modes". In: *Int J Numer Meth Eng.* 19(12): 1741–1752, 1983. DOI: 10.1002/nme.1620191202.

[112] Pian THH and Tong P. "Relations between Incompatible Displacement Model and Hybrid Stress Model". In: *Int J Numer Meth Eng.* 22(1): 173–181, 1986. DOI: 10.1002/nme.1620220112.

[113] Piltner R and Taylor RL. "A Quadrilateral Mixed Finite Element with Two Enhanced Strain Modes". In: *Int J Numer Meth Eng.* 38(11): 1783–1808, 1995. DOI: 10.1002/nme.1620381102.

[114] Piltner R and Taylor RL. "A Systematic Construction of B-bar Functions for Linear and Non-Linear Mixed-Enhanced Finite Elements for Plane Elasticity Problems". In: *Int J Numer Meth Eng.* 44(5): 615–639, 1999. DOI: 10.1002/(SICI)1097-0207(19990220)44:5<615::AID-NME518>3.0.CO;2-U.

[115] Puso MA. "A Highly Efficient Enhanced Assumed Strain Physically Stabilized Hexahedral Element". In: *Int J Numer Meth Eng.* 49(8): 1029–1064, 2000. DOI: 10.1002/1097-0207(20001120)49:8<1029::AID-NME990>3.0.CO;2-3.

[116] Rajendran S. “A Technique to Develop Mesh-Distortion Immune Finite Elements”. In: *Comput Methods Appl Mech Eng.* 199(17): 1044–1063, 2010. DOI: 10.1016/j.cma.2009.11.017.

[117] Rajendran S and Liew KM. “A Novel Unsymmetric 8-Node Plane Element Immune to Mesh Distortion under a Quadratic Displacement Field”. In: *Int J Numer Meth Eng.* 58(11): 1713–1748, 2003. DOI: 10.1002/nme.836.

[118] Rajendran S and Liew KM. “Completeness Requirements of Shape Functions for Higher Order Finite Elements”. In: *Struct Eng Mech.* 10(2): 93–110, 2000. DOI: 10.12989/sem.2000.10.2.093.

[119] Reddy BD. *Introductory Functional Analysis: With Applications to Boundary Value Problems and Finite Elements.* New York: Springer, 1998. DOI: 10.1007/978-1-4612-0575-3.

[120] Reddy BD and Simo JC. “Stability and Convergence of a Class of Enhanced Strain Methods”. In: *SIAM J Numer Anal.* 32(6): 1705–1728, 1995. DOI: 10.1137/0732077.

[121] Reese S. “Theorie und Numerik des Stabilitätsverhaltens hyperelastischer Festkörper”. PhD thesis. Darmstadt: TH Darmstadt, 1994.

[122] Reese S. “On a Consistent Hourglass Stabilization Technique to Treat Large Inelastic Deformations and Thermo-Mechanical Coupling in Plane Strain Problems”. In: *Int J Numer Meth Eng.* 57(8): 1095–1127, 2003. DOI: 10.1002/nme.719.

[123] Reese S. “On a Physically Stabilized One Point Finite Element Formulation for Three-Dimensional Finite Elasto-Plasticity”. In: *Comput Methods Appl Mech Eng.* 194(45): 4685–4715, 2005. DOI: 10.1016/j.cma.2004.12.012.

[124] Reese S and Wriggers P. “A Stabilization Technique to Avoid Hourglassing in Finite Elasticity”. In: *Int J Numer Meth Eng.* 48(1): 79–109, 2000. DOI: 10.1002/(SICI)1097-0207(20000510)48:1<79::AID-NME869>3.0.CO;2-D.

[125] Schmied C. “Hocheffiziente und robuste finite Elemente niederer Ansatzordnung zur Simulation großer Deformationen von Festkörpern”. PhD thesis. Karlsruhe: Karlsruhe Institute of Technology (KIT), 2017. DOI: 10.5445/IR/1000085307.

[126] Schröder J, Wriggers P, and Balzani D. “A New Mixed Finite Element Based on Different Approximations of the Minors of Deformation Tensors”. In: *Comput Methods Appl Mech Eng.* 200(49-52): 3583–3600, 2011. DOI: 10.1016/j.cma.2011.08.009.

[127] Schröder J et al. “A Selection of Benchmark Problems in Solid Mechanics and Applied Mathematics”. In: *Arch Comput Meth Eng.*, 2020. DOI: 10.1007/s11831-020-09477-3.

[128] Schwer LE, Key SW, Pučik TA, Pu TA, and Bindeman LP. “An Assessment of the LS-DYNA Hourglass Formulations via the 3D Patch Test”. In: Proc. 5th European LS-DYNA Users Conference. Birmingham, 2005. URL: https://www.dynalook.com/conferences/european-conf-2005/Schwer.pdf (visited on 2022-11-08).

[129] Simo JC. "A Framework for Finite Strain Elastoplasticity Based on Maximum Plastic Dissipation and the Multiplicative Decomposition. Part II: Computational Aspects". In: *Comput Methods Appl Mech Eng.* 68(1): 1–31, 1988. DOI: 10.1016/0045-7825(88)90104-1.

[130] Simo JC. "Algorithms for Static and Dynamic Multiplicative Plasticity That Preserve the Classical Return Mapping Schemes of the Infinitesimal Theory". In: *Comput Methods Appl Mech Eng.* 99(1): 61–112, 1992. DOI: 10.1016/0045-7825(92)90123-2.

[131] Simo JC and Armero F. "Geometrically Non-Linear Enhanced Strain Mixed Methods and the Method of Incompatible Modes". In: *Int J Numer Meth Eng.* 33(7): 1413–1449, 1992. DOI: 10.1002/nme.1620330705.

[132] Simo JC, Armero F, and Taylor RL. "Improved Versions of Assumed Enhanced Strain Tri-Linear Elements for 3D Finite Deformation Problems". In: *Comput Methods Appl Mech Eng.* 110(3-4): 359–386, 1993. DOI: 10.1016/0045-7825(93)90215-J.

[133] Simo JC and Hughes TJR. *Computational Inelasticity.* Vol. 7. Interdisciplinary Applied Mathematics. New York: Springer, 1998. DOI: 10.1007/b98904.

[134] Simo JC and Rifai MS. "A Class of Mixed Assumed Strain Methods and the Method of Incompatible Modes". In: *Int J Numer Meth Eng.* 29(8): 1595–1638, 1990. DOI: 10.1002/nme.1620290802.

[135] Simo JC, Taylor RL, and Pister KS. "Variational and Projection Methods for the Volume Constraint in Finite Deformation Elasto-Plasticity". In: *Comput Methods Appl Mech Eng.* 51(1): 177–208, 1985. DOI: 10.1016/0045-7825(85)90033-7.

[136] Sprenger W and Wagner W. "On the Formulation of Geometrically Nonlinear 3D-Rebar-elements Using the Enhanced Assumed Strain Method". In: *Eng Struct.* 21(3): 209–218, 1999. DOI: 10.1016/S0141-0296(97)00182-X.

[137] Stolarski H and Belytschko T. "Limitation Principles for Mixed Finite Elements Based on the Hu-Washizu Variational Formulation". In: *Comput Methods Appl Mech Eng.* 60(2): 195–216, 1987. DOI: 10.1016/0045-7825(87)90109-5.

[138] Sussman T and Bathe KJ. "Spurious Modes in Geometrically Nonlinear Small Displacement Finite Elements with Incompatible Modes". In: *Comput Struct.* 140: 14–22, 2014. DOI: 10.1016/j.compstruc.2014.04.004.

[139] Sze KY. "On Immunizing Five-Beta Hybrid-Stress Element Models from 'Trapezoidal Locking' in Practical Analyses". In: *Int J Numer Meth Eng.* 47(4): 907–920, 2000. DOI: 10.1002/(SICI)1097-0207(20000210)47:4<907::AID-NME808>3.0.CO;2-A.

[140] Sze KY, Chow CL, and Wanji C. "On Invariance of Isoparametric Hybrid/Mixed Elements". In: *Commun Numer Meth Eng.* 8(6): 385–406, 1992. DOI: 10.1002/cnm.1630080605.

[141] Taylor RL, Beresford PJ, and Wilson EL. “A Non-Conforming Element for Stress Analysis”. In: *Int J Numer Meth Eng.* 10(6): 1211–1219, 1976. DOI: 10.1002/nme.1620100602.

[142] Taylor RL, Simo JC, Zienkiewicz OC, and Chan ACH. “The Patch Test—a Condition for Assessing FEM Convergence”. In: *Int J Numer Meth Eng.* 22(1): 39–62, 1986. DOI: 10.1002/nme.1620220105.

[143] Ten Eyck A and Lew A. “An Adaptive Stabilization Strategy for Enhanced Strain Methods in Non-Linear Elasticity”. In: *Int J Numer Meth Eng.* 81(11): 1387–1416, 2010. DOI: 10.1002/nme.2734.

[144] Triantafyllidis N, Scherzinger WM, and Huang H. “Post-Bifurcation Equilibria in the Plane-Strain Test of a Hyperelastic Rectangular Block”. In: *Int J Solids Struct.* 44(11): 3700–3719, 2007. DOI: 10.1016/j.ijsolstr.2006.10.012.

[145] Viebahn N, Schröder J, and Wriggers P. “An Extension of Assumed Stress Finite Elements to a General Hyperelastic Framework”. In: *Adv Model and Simul in Eng Sci.* 6(9): 1–9, 2019. DOI: 10.1186/s40323-019-0133-z.

[146] Wagner W and Gruttmann F. “A Robust Non-Linear Mixed Hybrid Quadrilateral Shell Element”. In: *Int J Numer Meth Eng.* 64(5): 635–666, 2005. DOI: 10.1002/nme.1387.

[147] Wall WA, Bischoff M, and Ramm E. “A Deformation Dependent Stabilization Technique, Exemplified by EAS Elements at Large Strains”. In: *Comput Methods Appl Mech Eng.* 188(4): 859–871, 2000. DOI: 10.1016/S0045-7825(99)00365-5.

[148] Washizu K. *Variational Methods in Elasticity and Plasticity.* 3rd ed. Oxford: Pergamon Press, 1982.

[149] Wilson EL, Taylor RL, Doherty WP, and Ghaboussi J. “Incompatible Displacement Models”. In: *Numerical and Computer Methods in Structural Mechanics.* Ed. by Fenves SJ, Perrone N, and Robinson AR. New York: Elsevier, 1973, pp. 43–57. DOI: 10.1016/b978-0-12-253250-4.50008-7.

[150] Wisniewski K and Turska E. “Improved Four-Node Hellinger-Reissner Elements Based on Skew Coordinates”. In: *Int J Numer Meth Eng.* 76(6): 798–836, 2008. DOI: 10.1002/nme.2343.

[151] Wisniewski K and Turska E. “Improved 4-Node Hu-Washizu Elements Based on Skew Coordinates”. In: *Comput Struct.* 87(7): 407–424, 2009. DOI: 10.1016/j.compstruc.2009.01.011.

[152] Wisniewski K, Wagner W, Turska E, and Gruttmann F. “Four-Node Hu-Washizu Elements Based on Skew Coordinates and Contravariant Assumed Strain”. In: *Comput Struct.* 88(21): 1278–1284, 2010. DOI: 10.1016/j.compstruc.2010.07.008.

[153] Wriggers P and Simo JC. “A General Procedure for the Direct Computation of Turning and Bifurcation Points”. In: *Int J Numer Meth Eng.* 30(1): 155–176, 1990. DOI: 10.1002/nme.1620300110.

[154] Wriggers P. *Nonlinear Finite Element Methods.* Berlin: Springer, 2008. DOI: 10.1007/978-3-540-71001-1.

[155] Wriggers P and Korelc J. "On Enhanced Strain Methods for Small and Finite Deformations of Solids". In: *Comput Mech.* 18(6): 413–428, 1996. DOI: 10.1007/BF00350250.

[156] Wriggers P and Reese S. "A Note on Enhanced Strain Methods for Large Deformations". In: *Comput Methods Appl Mech Eng.* 135(3-4): 201–209, 1996. DOI: 10.1016/0045-7825(96)01037-7.

[157] Wu CC and Cheung YK. "On Optimization Approaches of Hybrid Stress Elements". In: *Finite Elem Anal Des.* 21(1): 111–128, 1995. DOI: 10.1016/0168-874X(95)00023-0.

[158] Xie Q, Sze KY, and Zhou YX. "Modified and Trefftz Unsymmetric Finite Element Models". In: *Int J Mech Mater Des.* 12(1): 53–70, 2016. DOI: 10.1007/s10999-014-9289-3.

[159] Yuan KY, Huang YS, and Pian THH. "New Strategy for Assumed Stresses for 4-Node Hybrid Stress Membrane Element". In: *Int J Numer Meth Eng.* 36(10): 1747–1763, 1993. DOI: 10.1002/nme.1620361009.

[160] Zavarise G, De Lorenzis L, and Taylor RL. "A Non-Consistent Start-up Procedure for Contact Problems with Large Load-Steps". In: *Comput Methods Appl Mech Eng.* 205–208: 91–109, 2012. DOI: 10.1016/j.cma.2011.02.018.

[161] Zhou PL, Cen S, Huang JB, Li CF, and Zhang Q. "An Unsymmetric 8-Node Hexahedral Element with High Distortion Tolerance". In: *Int J Numer Meth Eng.* 109(8): 1130–1158, 2017. DOI: 10.1002/nme.5318.

[162] Zienkiewicz OC, Taylor RL, and Zhu J. *The Finite Element Method. Vol. 1: Its Basis and Fundamentals.* 6th ed. Amsterdam: Elsevier Butterworth-Heinemann, 2010. DOI: 10.1016/C2009-0-24909-9.

Schriftenreihe des Instituts für Mechanik
ISSN 2363-4936

Hrsg. von: Prof. Dr.-Ing. habil. Peter Betsch, Prof. Dr.-Ing. habil. Thomas Seelig

Band 1 Marlon Franke
Discretisation techniques for large deformation computational contact elastodynamics.
ISBN 978-3-7315-0278-4

Band 2 Philipp Hempel
Constitutive modeling of amorphous thermoplastic polymers with special emphasis on manufacturing processes.
ISBN 978-3-7315-0550-1

Band 3 Martin Helbig
Mehrskalenmodellierung von Schädigung in gummimodifizierten thermoplastischen Kunststoffen.
ISBN 978-3-7315-0571-6

Band 4 Maik Dittmann
Isogeometric analysis and hierarchical refinement for multi-field contact problems.
ISBN 978-3-7315-0616-4

Band 5 Yinping Yang
Numerical methods for the inverse dynamics simulation of underactuated mechanical systems.
ISBN 978-3-7315-0626-3

Band 6 Alexander Janz
Structure-preserving space-time discretization in a mixed framework for multi-field problems in large strain elasticity.
ISBN 978-3-7315-0926-4

Band 7 Mark Georg Schiebl
Thermodynamically consistent space-time discretization of non-isothermal mechanical systems in the framework of GENERIC.
ISBN 978-3-7315-1117-5

Band 8 Jonas Hund
Characterisation and modelling of PC/ABS blends.
ISBN 978-3-7315-1157-1

Band 9 Jonathan Schulte
Multi-field modeling and simulation of fiber-reinforced polymers.
ISBN 978-3-7315-1251-6

Band 10 Robin Pfefferkorn
EAS Elements for Solid Mechanics – Mesh Distortion Insensitive and Hourglassing-Free Formulations with Increased Robustness.
ISBN 978-3-7315-1271-4